Reactive scattering for $H^- + H_2$ and $H^+ + H_2$ and its isotopologues: Classical versus Quantum investigation

DISSERTATION

zur Erlangung des Grades eines Doktors

der Naturwissenschaften

vorgelegt von

M. Sc. Dequan Wang

geb. am 25 April 1975 in Jilin, China

eingereicht bei der Naturwissenschaftlich-Technischen Fakultät

der Universität Siegen
Siegen 2012

Bibliographic information published by the Deutsche Nationalbibliothek

The Deutsche Nationalbibliothek lists this publication in the Deutsche
Nationalbibliografie; detailed bibliographic data are available
on the Internet at http://dnb.d-nb.de .

ISBN 978-3-8325-4295-5

Logos Verlag Berlin GmbH
Comeniushof, Gubener Str. 47,
10243 Berlin
Tel.: +49 (0)30 42 85 10 90
Fax: +49 (0)30 42 85 10 92
INTERNET: http://www.logos-verlag.de

Gutachter: 1. Professor Dr. Ralph Jaquet
                     2. Professor Dr. Thomas Lenzer
                     3. Professor Dr. Michael Schmittel
                     4. PD Dr. Kawon Oum

Eingereicht am: 30 Nov. 2012
Tag der mündlichen Prüfung: 18 Jan. 2013

Acknowledgements

First of all, I would like to express my most memorable appreciation to my supervisor, Prof. Dr. Ralph Jaquet, for giving me the opportunity to work in his group at the Department of Chemistry, University of Siegen, Germany. His broad knowledge in chemistry and infinite wisdom enlighten me very much to find solutions when I meet problems during my Ph.D work. I would also like to thank him for his encouragement and help during my stay in Siegen.

My heartfelt thanks to Prof. Dr. Thomas Lenzer and PD Dr. Kawon Oum for help when I prepare my Ph.D thesis.

Additionally, I would like to give special thanks to Prof. Dr. Xuri Huang for recommending me studying in Siegen University.

I also thank the secretary Mrs. Marie Luise Kleinschmidt for her timely help, and Prof. Dr. Bo Song, Dr. Qinghai Shu, MSc. Simon Haas, Dr. Mykhaylo Khoma, and MSc. Kun Chen for they help during my stay in Siegen. Thanks all of my friends who always share my failures and successes.

There are no words that I can express my gratitude to my father, my brothers and my sisters for their endless supports and love.

In particularly, I would like to thank my lovely wife Lijuan Zhou for her long-term understanding and support.

This Project was sponsored by the Scientic Research Foundation for the Returned Overseas Chinese Scholars, State Education Ministry. For the financial support I am thankful to the China Scholarship Council, the Deutsche Forschungsgemeinschaft and to the University of Siegen.

Summary

In the present doctoral thesis, the reactive scattering for $H^- + H_2$ and $H^+ + H_2$ and its iso-topologues were investigated using different methods to solve the equations describing classical and quantum mechanics. The studies aimed at providing insights into elementary reactions, and may even go beyond these to more complex chemical reactions. The main results in this dissertation can be summarized as follows:

In **Chapter 2** the equations solving problems in quasi-classical mechanics were described, which led to the definition of energy dependent reaction probabilities $P_r(E_{rel}, v, j) = \frac{N_r}{N_{tot}}$ and reaction cross sections $\sigma_R = \pi P_r(E_{rel}, v, j) b_{max}^2$.

The formalism for time-dependent methods for the investigation of scattering processes was presented in **Chapter 3**. In this section we discussed how to use the time-dependent quantum wavepacket method to study the A-BC system. The dependence of the reaction probabilities $P_{reac}^J(E)$ on the total angular momenta J was calculated to obtain information about the integral reactive cross section $\sigma^{tot}(E) = \frac{\pi}{k_{vj}^2} \sum_J (2J + 1) P_{reac}^J(E)$.

The potential energy surfaces (PESs) for H_3^+ and H_3^- were described in **Chapter 4**. For the H_3^+ system, a cut through the potential energy surface (PES) in the asymptotic region was presented. For the H_3^- system three available *ab initio* potential energy surfaces have been used in the applications: a) Stärck and Meyer (SM-PES), b) Panda and Sathyamurthy (PS-PES), and c) Ayouz et al. (AY-PES). The differences in the PESs were investigated.

In the beginning of **Chapter 5** the $H^+ + H_2(v{=}0{-}5,\ j{=}0)$ collision was investigated non-adiabatically. By comparison of the reaction probabilities using adiabatic and non-adiabatic representations of the potential energy surfaces, it was found that, at low collision energies, the reaction preferentially occurs adiabatically, but at higher collision energies non-adiabatic effects have to be taken into account.

Reaction probabilities and reaction cross sections for the collision H^- with H_2 and its isotopologues using quasi-classical trajectories and quantum wavepackets were presented in the main part of **Chapter 5**. It was found that, at low collision energies, the reaction probabilities using SM-PES and AY-PES are very similar. The reaction probabilities based on the PS-PES are lower than those based on the SM-PES and AY-PES. At lower collision energies the reaction cross sections calculated with SM-PES are higher than those calculated with PS-PES. The reaction cross sections investigated with quasi-classical trajectories are higher than those calculated with quantum wavepackets (using the same potential).

The last section of **Chapter 5** showed results for the collision of H^- and D^- with HD. The total reaction probabilities, the reaction cross sections, and the product ratios were determined using quasi-classical trajectories. One can learn from these calculations that for the H^- + HD(v=0–1, j=0) reaction and low collision energies, the main product are H_2 + D^-. At high collision energies, the product channel HD + H^- is slightly dominant. For the collision of D^- with HD and low collision energies the product channel HD + D^- is strongly favored, but in the high collision energy range, the product channel D_2 + H^- dominates.

Zusammenfassung

In der vorliegenden Doktorarbeit wurde die reaktive Streuung für $H^- + H_2$ und $H^+ + H_2$, sowie dessen Isotopologe untersucht, indem die Gleichungen, welche die klassische Mechanik und die Quantenmechanik beschreiben, mit verschiedenen Methoden gelöst wurden. Das Ziel dieser Studien ist, neue Einblicke in elementare Reaktionen zu gewinnen und diese darüber hinaus, auf komplexere chemische Reaktionen zu übertragen. Die Hauptergebnisse dieser Dissertation können wie folgt zusammengefasst werden:

In **Kapitel 2** werden die Gleichungen zum Lösen von Problemen mit Hilfe der quasi-klassischen Mechanik beschrieben, welche zu der Definition der energieabhängigen Reaktionswahrscheinlichkeiten $P_r(E_{rel}, v, j) = \frac{N_r}{N_{tot}}$ und der Reaktionsquerschnitte $\sigma_R = \pi P_r(E_{rel}, v, j) b_{max}^2$ führen.

Der Formalismus für zeitabhängige Methoden zur Untersuchung von Streuprozessen wird in **Kapitel 3** präsentiert. In diesem Abschnitt erörtern wir, wie die A–BC Systeme mit Hilfe der zeitabhängigen, quantenmechanischen Wellenpaketmethode studiert werden können. Die Abhängigkeit der Reaktionswahrscheinlichkeiten $P_{reac}^J(E)$ vom Gesamtdrehimpuls J wird berechnet, um Informationen über den integralen Reaktionsquerschnitt $\sigma^{tot}(E) = \frac{\pi}{k_{vj}^2} \sum_J (2J + 1) P_{reac}^J(E)$ zu erhalten.

Die potentielle Energieflächen (PEF) für H_3^+ und H_3^- werden in **Kapitel 4** beschrieben. Für das H_3^+-System wird ein Schnitt durch die PEF in der asymptotischen Region präsentiert. Für das H_3^--System hingegen werden die drei verfügbaren potentiellen Energieflächen von Stärck und Meyer (SM-PEF), Panda und Sathyamurthy (PS-PEF), sowie Ayouz et al. (AY-PEF) verwendet. Die Unterschiede in den PEFn, die auf *ab initio* Energien basieren, werden untersucht.

Zu Beginn von **Kapitel 5** wird der (H^+ + H_2(v=0–5,j=0))-Stoß nicht-adiabatisch untersucht. Beim Vergleich der Reaktionswahrscheinlichkeiten die mit Hilfe adiabatischer und nicht-adiabatischer Darstellung der potentiellen Energiefläche berechnet wurde stellt man fest, dass bei niedrigen Stoßenergien die Reaktion einen adiabatischen Prozess bevorzugt, während bei höheren Stoßenergien auch nicht-adiabatische Effekte in Betracht gezogen werden müssen.

Die Reaktionswahrscheinlichkeiten und Reaktionsquerschnitte für den Stoß von H^- mit H_2 und seine Isotopologe unter Verwendung von quasi-klassischen Trajektorien und quantenmechanischen Wellenpaketen werden im Hauptteil von **Kapitel 5** dargestellt. Es wurde herausgefunden, dass bei niedrigen Stoßenergien die Reaktionswahrscheinlichkeiten für SM-PEF und AY-PEF sehr ähnlich sind. Die Reaktionswahrscheinlichkeiten basierend auf PS-PEF sind hingegen niedriger als die auf SM-PEF und AY-PEF basierenden. Bei niedrigeren Stoßenergien sind die mit SM-PEF berechneten Reaktionsquerschnitte größer als die mit PS-PEF ermittelten. Die Reaktionsquerschnitte, welche mit quasi-klassischen Trajektorien untersucht wurden, sind höher als die mit quantenmechanischen Wellenpaketen berechneten (unter Verwendung des gleichen Potentials).

Der letzte Abschnitt von **Kapitel 5** zeigt Ergebnisse für die Stöße von H^- und D^- mit HD. Die Gesamtreaktionswahrscheinlichkeiten, die Reaktionsquerschnitte und die Produktverhältnisse wurden unter Verwendung von quasi-klassischen Trajektorien bestimmt. Aus diesen Berechnungen kann man lernen, dass bei niedrigen Stoßenergien für die H^- + HD(v=0–1, j=0) Reaktion, H_2 + D^- das Hauptprodukt darstellt. Bei hohen Stoßenergien dominiert der (HD + H^-)–Kanal etwas. Für den Stoß von D^- mit HD wird bei niedrigen Stoßenergien der Produktkanal HD + D^- stark favorisiert, während im hohen Stoßenergiebereich der (D_2 + H^-)–Produktkanal dominiert.

List of Symbols and Abbreviations

PES	Potential Energy Surface
WP	WavePacket
SF	Space-Fixed (coordinate system)
BF	Body-Fixed (coordinate system)
DVR	Discrete Vabiable Representation
FFT	Fast Fourier Transform
R, r, θ	Jacobi coordinates
RC	Reactant Jacobi Coordinates
PC	Product Jacobi Coordinates
v	vibrational quantum number of the diatomic (BC) molecule
j	rotational quantum number of the diatomic (BC) molecule
J	the total angular momentum quantum number of an A-BC system

Atomic units

The most important atomic units are summarised in the table below:[1]

Obervable	Atomic unit
Energy	27.211835 eV (1 hartree - E_h)
Length	$0.5291772 \times 10^{-10}$ (1 bohr - a_0)
Time	$2.4188843 \times 10^{-17}$ s
Mass	9.109382×10^{31} kg

[1]Latest (2010) values of the constants 'CODATA Internationally recommended values of the Fundamental Physical Constants' is available at http://physics.nist.gov/cuu/Constants/

Contents

1 Introduction

Collisions involving hydrogen atoms, molecules, and their positive (H^+, H_2^+) and negative (H^-) ions play an important role in chemistry and the evolution of neutral or negatively-charged hydrogen plasma such as laboratory hydrogen plasma, the interstellar medium (ISM), the atmospheres of the sun and other stars [1] as well as in the atmosphere of the Earth.

The first calculation of the potential energy surface for the H_3 molecule and the H_3^+ ion was performed by Hirschfelder et al. [2, 3, 4]. The activation energy for the reaction $H + H_2 \rightarrow H_2$ + H is 13.63 kcal/mol using the Heitler-London plus polar states. The experimental reaction barrier is 5.5 kcal/mol [5].

In the 1910s, Thomson [6] discovered the H_3^+ ion as a mass/charge $= 3m_p/e$ ray in the spectrum which was called the mass spectrum in today's language produced by an electrical discharge through hydrogen gas. Five years later, Dempster [7] (1916) confirmed this observation using electron-beam excitation of hydrogen. There is experimental evidence that the H_3^+ ion is formed whenever H_2 is ionized at any but the lowest pressures [8]. The main processes which can produce H_3^+ ions are the following secondary processes:

$$\mathbf{H_2^+ + H_2 \rightarrow H_3^+ + H,} \tag{1}$$

$$\mathbf{H^+ + H_2 \rightarrow H_3^+,} \tag{2}$$

$$\mathbf{H + H_2^+ \rightarrow H_3^+,} \tag{3}$$

in which the first process is usually the most important. In any event, the reaction $H_2^+ + H_2 \rightarrow H_3^+ + H$ should happen with a small amount of kinetic energy, and it is probable that it can occur without any relative kinetic energy. Hirschfelder et al. [3] have calculated the potential energies for linear symmetrical configurations of H_3^+. The reasons for why H_3 should be linear and H_3^+ should be triangular has been discussed by Coulson [9]. There are two electrons in the H_3^+ ion, and so, in the ground state, only one type of orbital is filled; the two electrons will have the same spatial wave function but opposite spins. One chooses the configuration of the nuclei which gives the orbital the greatest bonding. If the three H atoms (see Fig. 1) are arranged in a linear structure as $[H_a-H_b-H_c]^+$, then the lowest orbital will only represent resonance between the atoms H_a and H_b and between the atoms H_b and H_c. The resonance between the atoms H_a and H_c is too small to make any appreciable contribution to bonding. But, in the triangular model, all three resonances contribute, and we may therefore expect greater bonding. In the neutral molecule H_3, however, a new type of orbital has to be introduced with a node along the dotted line (see Fig. 1); this provides H_a–H_b and H_b–H_c bonding but H_a–H_c repulsion and therefore a definite tendency for this molecule is open into a flat triangle or a straight line.

Because of its fundamental importance, the H_3^+ ion has been the subject of theoretical, experimental, and astrophysical investigations in recent decades [10, 11, 12], and a discussion

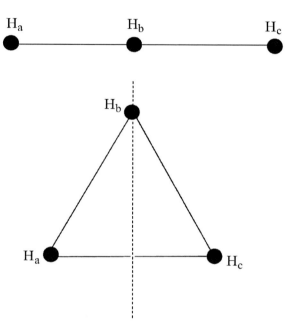

Figure 1: Linear and triangle model of H_3 or H_3^+.

meeting entitled "Physics, Chemistry and Astronomy of H_3^+" has been held by the Royal Society three times in 2000, 2006, and 2012. As discussed above, the H_3^+ ion has an equilateral triangular equilibrium geometry in its ground state. The linear symmetric configuration, 14299 cm^{-1} above the equilateral minimum, is a saddle point. The dissociation limit into $H^+ + H_2$ lies at 37170 cm^{-1} above the minimum. The H_3^+ ion has no permanent dipole moment in the equilibrium structure since it is symmetrical. There are two vibrational modes in the H_3^+ ion: the totally symmetric stretch mode ν_1 and the doubly degenerate bending mode ν_2, which is the only one that is infrared active. Oka [13] found the first infrared $\nu_2 \rightarrow 0$ lines in the laboratory in 1980. Since then, numerous laboratory experiments have led to the detection and identification of nearly 900 lines, which were compiled in 2001 [14], all of which involve energies below the barrier to the linear configuration. Recently, some studies [15, 16] have been done considering transitions to the states slightly above the linear barrier. Other experiments [17, 18, 19] have concentrated on the predissociation of H_3^+ near the upper bound of the spectrum within 1100 cm^{-1} of the dissociation limit of $H^+ + H_2$.

Nearly at the same time, the H_3^+ ion has been researched by astrophysicists in the universe through its infrared spectra. H_3^+ ion was fist identified by Drossart et al. [20] in the Jovian atmosphere in 1989, and then it was found by Geballe and Oka [21] and McCall et al. [22] in

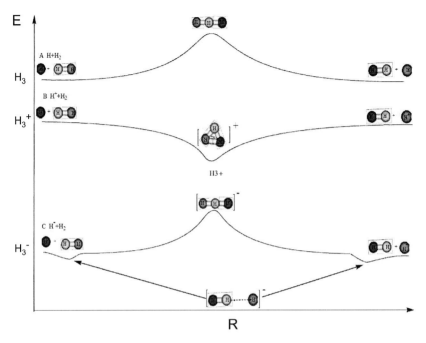

Figure 2: Energy of an H (H^+, or H^-) atom (ion) approaching H_2 as a function of its separation R.

dense interstellar clouds.

A large number of theoretical studies were essential in the identification and understanding of the H_3^+ ion. The investigation of the electronic potential energy surface (PES) is the first essential ingredient. There are three kinds of PESs for the H_3^+ ion. The first family is the one containing local *ab initio* PESs. There are five PESs [23, 24, 25, 26, 27] of this family that are available in the literature. The MBB [23] potential is an example of such a PES. It was obtained from a full CI investigation at 69 points with a maximum energy of 25000 cm^{-1} above the minimum. This PES is not accurate at the dissociation region, so it cannot be used to compute highly excited bound states or to study dynamic processes such as the $H^+ + H_2$ inelastic charge exchange or reactive collisions. The RKJK [24] PES is of higher accuracy. It relies on the highly accurate CI-R12-method, namely, a configuration interaction method explicitly including an r_{12} linear term in the wave function. Even greater accuracy was obtained in 1998 (JCKR [26]) using correlated Gaussian functions and including adiabatic and relativistic corrections. An accuracy as high as 0.02 cm^{-1} has been claimed. In 2002, the JCKR PES was improved by Jaquet [27]. The local region of the JCKR potential was

3

extended to higher energies by a further 130 *ab initio* points obtained with the CI-R12 method in addition to the initial 69 points obtained with correlated Gaussian functions. The second family of PESs is the semi-empirical one obtained by iterative adjustment of the potential until good agreement with the experimental spectroscopic data is obtained. The DMT-PES [28] is an example of this family: it was obtained by the adjustment to 243 experimental energy levels with a standard deviation of 0.053 cm^{-1}. The third family of PESs includes *global* potential energy surfaces. This kind of PES relies on *ab initio* calculations, usually performed on a large number of points covering the entire configuration space. The third family of PESs provides slightly lower accuracy than local PESs for the computational infrared spectra, but they are also useful to compute highly excited rovibrational states and dynamic processes above the dissociation threshold. Six PESs [29, 30, 31, 32, 33, 39] belong to this family. The sixth one is the VAV-PES [39], which is a global PES for singlet H_3^+ based on the method proposed by Varandas [40].

For the singlet state of the H_3^+ ion there are two PESs available in the literature that can be used to study non-adiabatic processes. The first is the Preston and Tully surface [34] obtained in 1971. The second one is the KBNN [35] PES, which gives a better description of the avoided crossing.

Under the conditions in which H_3^+ is present, additional $H_3^+(H_2)_n$ complexes are also stable [36]. Among these, the ion H_5^+ plays a special role. While H_3^+ is the prototype of a 3-centre-2-electron bond, H_5^+ represents a 5-centre-4-electron bond. While the bond between a proton and a H_2 molecule is particularly strong ($D_e = 169$ $mE_h = 444$ kJ mol^{-1}), the gain upon attaching H_3^+ to another H_2 is relatively small ($D_e = 13.7$ $mE_h = 35.9$ kJ mol$^{-1} = 3007$ cm^{-1}). The chemical reaction $H_3^+ + H_2 \rightarrow H_2 + H_3^+$ had been studied in a hollow cathode plasma cell by Crabtree et al. [37]. The ratio of the rates of the proton hop (k^H) and hydrogen exchange (k^E) reactions $\alpha \equiv k^H/k^E$ has been found to decrease from 1.6 ± 0.1 at 350 K to 0.5 ± 0.1 at 135 K. Gómez-Carrasco et al. [38] investigated the $H_3^+ + H_2 \rightarrow H_2 + H_3^+$ using quasi-classical trajectories. The α value was in good agreement with the experimental results by Crabtree et al. [37].

The negative ion H_3^- has been investigated rather intensively experimentally and theoretically in the past 50 years [42, 43, 44, 45, 46, 47, 48, 50, 51, 52, 53, 54, 55, 56, 57, 58, 59, 60, 61, 62, 63, 64, 65, 66, 67, 68, 69, 70, 71, 72, 73, 74, 75, 76]. The potential energy surface of H_3^- is very similar to that of H_3, including the anisotropy, barrier height, and other characteristics. The reason is that the additional electron in the H_3^- ion is located at a rather large distance from all nuclei and, as a consequence, the main interaction is given by the three nuclei and the three "inner" electrons (Fig. 2). There are differences between the H_3^- ion and the H_3 molecule. The first difference is the charge-induced dipole interaction that leads to a shallow well in the H_3^- molecule at greater H^-–H_2 distances. The PES by Stärck and Meyer [50] shows a well depth of 0.048 eV at $R = 6.183$ Bohr and $r = 1.413$ Bohr. The second significant difference between the neutral and the negative ion system is that the fourth electron gives rise to additional reaction channels involving electron detachment, i.e., the H_3^- PES includes the following additional reaction channel:

$$H^- + H_2 \to H + H_2 + e^-.$$

The first study of the H_3^- ion was performed by Stevenson and Hirschfelder [41] in 1937. They suggested the existence of bound states of the H_3^- ion in a linear configuration. The H_3^- molecular ion was discovered experimentally by Hurley [42] in 1974. Aberth et al. [43] (1975) reported the discovery of H_3^-, H_2D^-, HD_2^- and D_3^- ions, and claimed that these ions appear stable. Subsequently, H_3^- ions have been investigated in many experiments [44, 45, 46, 47, 48]. Collisions between H^- with H_2 should play a role in processes in tokamaks, which is a device that uses a magnetic field to confine plasma in the shape of a torus [49]. Although experiments started as early as 1974, it was only in the 1990s [50, 51] that the theory became precise enough to confirm the stability of H_3^- bound states. The H_3^- molecule is composed of three identical nuclei, described in principle within the CNPI (complete nuclear permutation inversion) group D_{3h} [52].

At present, there are three accurate potential energy surfaces (PESs) of H_3^- available: the PESs by Stärck and Meyer [50], by Panda and Sathyamurthy [53], and by Ayouz et al. [54]. In addition, Belyaev, Tiukanov and others [55, 56, 57, 58, 59, 60] have obtained PESs of excited electronic states of H_3^- and their non-Born-Oppenheimer couplings with the ground state. The excited electronic states are unstable with respect to electron autodetachment.

The collisions between H_2 and H^-, and of their isotopologues, have been studied in a number of laboratory experiments since the 1950s [61, 62, 63, 64, 65, 66, 67]. Muschlitz et al. [61, 62] improved an apparatus for the production of beams of negative ions and the measurement of the elastic and inelastic scattering of negative ions in gases at low pressures. They found that the inelastic cross section increases from 1 cm^{-1} at 7 eV to 8.7 cm^{-1} at 395 eV. Absolute total cross sections for electron detachment were measured for collisions between H^- and D^- ions with H_2, D_2, and HD by Huq et al. [63] in 1983. A crossed beam study of the rearrangement reaction $H^- + D_2(v{=}0) \to HD(v') + D^-$ in the collision energy range $E_{rel} = 0.3$–3 eV was reported by Zimmer and Linder [64, 65]. These reports showed that the reaction has a threshold at $E_{rel} = 0.42 \pm 0.12$ eV and the cross section rises to a maximum of 2×10^{-16} cm^2 at 1.5 eV and then rapidly decreases again. These studies showed that theoretical studies for this reaction are needed. With the combined efforts of theory and experiment, the $H^- + H_2/D_2$ system was established as a benchmark system for the present class of processes, i.e., reactive processes of negative ions including detachment channels. Crossed-beam measurements of rotationally inelastic scattering of H^- ions from H_2 were reported for collision energies in the energy range $E_{rel} = 1.66$–2.79 eV by Müller et al. [66]. The experiment showed that the rotational levels j' = 5, 7, 9 are dominantly excited. This striking behavior can be understood qualitatively from the properties of the PES of H_3^-. Integral cross sections for rotationally inelastic scattering has been estimated and compared with those for reactive scattering and electron detachment, which are competing processes in this energy range. Haufler et al. [67] determined the absolute integral and differential cross sections for the two reactions $H^- + D_2 \to D^- + HD$ and $D^- + H_2 \to H^- + HD$ in 1997.

5

Collisions between H_2 and H^- have been investigated by quantum mechanical methods [68, 69, 70, 71, 72, 73, 74, 75, 76] which started around the 1990s. Ayouz et al. [77] investigated the possibility of forming H_3^- by radiative association (RA) of H_2 and H^- in low temperature (< 150 K) environments.

Up to now, there have been no accurate classical trajectory studies for high vibrational states. The main purpose of the present work was to investigate chemical reactions using time-dependent quantum wave packets and quasi-classical trajectories for systems involving three atoms. Time-dependent quantum mechanical wave packets have become a practical tool in studying a wide variety of molecular processes due to their ease in implementation. The first time-dependent quantum mechanical methods to solve the time-dependent Schödinger equation were used for a collinear exchange reaction of the type A + BC → AB + C (1959, Mazur and Rubin [78]). Later this method was improved by many theorists. The fast Fourier transform (FFT) method was introduced by Kosloff and Kosloff [79]. This method is used to compute the action of the kinetic energy as part of the Hamiltonian on the wave function. The FFT method is an important development in the area of time-dependent quantum mechanics (TDQM). Light and coworkers [80, 81] developed the discrete variable representation (DVR), which is used for calculating matrix elements. In 1984, a global propagation scheme, based on the Chebyshev polynomial expansion of the evolution operator, was introduced by Tal-Ezer and Kosloff [82]. This propagation scheme was improved by Mandelshtam and Taylor in 1995 [83, 84]. A new algorithm for calculating only the real part of the wave function was introduced by Gray and Balint-Kurti [85].

Whenever the dimensionality or the number of open channels becomes larger or the system which should be studied becomes larger, quantum mechanical methods rapidly become intractable. In such cases, a classical or quasi-classical approach should be used. The quasi-classical method has been reviewed many times [86, 87, 88, 89, 90, 91, 92, 93, 94]. In the case of classical methods, the nuclei are assumed to move classically on an adiabatic electronic potential energy surface (PES). If the initial states of the collision are taken to be quantized, the procedure is termed a quasi-classical trajectory method. Classical trajectories are frequently used to investigate homogeneous gas-phase bimolecular processes. The quantities of interest in such studies commonly include cross sections, thermal rates, activation parameters, product energy partitioning, angular scattering, and the mechanism. Trajectories are also often used to examine unimolecular dissociation reactions [95] and gas-surface phenomena [96]. In our studies, the collisions between H_2 and H^- and its isotopologues have been fully investigated with quasi-classical trajectories and then compared with quantum methods; the details are provided in Chapter 5.

Molecular dynamics (MD), the numerical integration of the classical equations of motion describes the motions of atoms interacting on a multidimensional potential energy surface [97]. MD have proven to be extremely valuable for elucidating the dynamics of a very wide range of elementary chemical processes including gas phase collisions of small molecules, reactions in liquids, gas-surface interactions, materials properties, and protein dynamics. But even when accurate interaction potentials can be obtained, there may be significant limitations imposed by the fundamental assumptions on which the molecular dynamics method is founded.

The first approximation is the Born-Oppenheimer (adiabatic approximation), in which it is supposed that the nuclei, being so much heavier than an electron, move relatively slowly and may be treated as stationary while the electrons move in their field. We can therefore think of the nuclei as being fixed at arbitrary locations, and then solve the Schrödinger equation for the wave function of the electrons alone. So the motion of the nuclei can be assumed to be governed by a single adiabatic potential energy surface. The approximation is quite good for ground-state molecules. However, it often fails to describe reaction involving electronic transitions, e.g., photochemistry and laser-induced chemistry, electron transfer, reactions on metal surfaces, electronic energy transfer, non-radiative transitions, and ion-molecule reactions. The second fundamental assumption of the molecular dynamics method is that nuclei evolve according to classical mechanics. This is inadequate in many cases.

In chemistry, the terms adiabatic and non-adiabatic imply a separation of coordinates into two classes, which we will refer to throughout as fast and slow. Indeed, if all coordinates are treated on an equal footing, the meanings of adiabatic and non-adiabatic become obscured. Non-adiabatic transitions are formally indistinguishable from any other transitions between quantum states. The treatment of all degrees of freedom on an equal footing is a good goal, which means a full quantum mechanical description is required, so this method is only suitable for very small systems. For large systems fast and slow coordinates should be designated from the outset. For convenience, most of the "fast" coordinates are defined for electrons and the "slow" ones for nuclei. However, distinguishing fast and slow nuclear degrees of freedom is also considered.

Electronically non-adiabatic behavior is observed in a great number of fundamental molecular processes [98]. These include:

1. Electronic energy transfer (the asterisk denotes electronic excitation)
$$A^* + B \rightarrow A + B^* .$$

2. Charge transfer
$$A^+ + B \rightarrow A + B^+ .$$

3. Quenching of electronic excitation
$$A^* + B \rightarrow A + B^\dagger,$$
where the dagger denotes internal (vibrational and rotational) excitation of molecule B.

4. Chemical reactions
$$A + B \rightarrow C + D.$$

Although some chemical reactions can be described within the adiabatic hypothesis, others cannot. Reactions involving electronically excited reactants or products are likely to exhibit non-adiabatic transitions because of the expected proximity of neighboring excited state potential energy surfaces. Similarly, ion-molecule reactions are frequently non-adiabatic due to the possibility of charge transfer. But, electronic transitions are common even in reactions of ground state species at room temperature. In some cases, the role of non-adiabatic coupling is readily apparent. Examples are spin forbidden reactions and reactions that are accompanied by excitation transfer. There are some other more subtle cases where non-adiabatic effects dramatically alter the chemical forces that determine the course of a reaction, but do not reveal themselves explicitly in the products. For all these reasons, the non-adiabatic methods are very important in theoretical chemistry.

The characteristics of the electronic structure that determine the reaction mechanisms underlying the simplest atom-ion-molecule collision system of H_3^+ and its isotopologues are not simple [36, 99]. The ground electronic state $^1A'$ asymptotically and adiabatically correlates with the reagent/product channel of $H^+ + H_2$, while the lowest excited $^1A'$ state correlates with $H + H_2^+$. Electronic structure studies have also revealed that, in the entrance and the exit regions far from the H_3^+ equilibrium geometry, avoided crossings exist between the ground $^1A'$ potential energy surface and the lowest excited $^1A'$ surface. These avoided crossing regions play a very important role in $H + H_2^+$ collision dynamics due to effective electronically non-adiabatic transitions, which open a pathway for charge transfer to form the molecular ion. Thus, reactive and nonreactive chemical processes below the dissociation energy of H_2^+ can take place with or without charge transfer in the H_3^+ system, depending on whether non-adiabatic transitions are involved or not. A three-dimensional "trajectory surface hopping" treatment of the reaction H^+ with D_2 at a collision energy of 4 eV was first reported by Tully and Preston [100]. In this report, the reaction cross sections were 0.33, 0.21, and 0.37 $Å^2$ for the products D^+, HD^+, and D_2^+. These results were in good agreement with the results of Holliday et al. [101]. Furthermore, the predicted value of 0.56 for the HD^+/D_2^+ ratio was in very good agreement with the value of 0.60 reported by Krenos and Wolfgang [102]. Since the 1990s, the non-adiabatic charge transfer reaction of $H^+ + H_2$ and its isotopic variants has stimulated quasi-classical trajectory and quantum mechanical calculations [103, 104, 105, 106, 107, 108, 109, 110, 111, 112, 113, 114, 115]. Driven by non-adiabatic couplings between different electronic states, charge transfer is the key in both gas phase and solvent reactions, and also plays a crucial role in many biological processes. Hence, investigations into non-adiabatic effects in the $H^+ + H_2$ reaction can provide insights into the associated charge transfer processes.

The outline of the present work is as follows. The second chapter introduces the quasi-classical trajectory method. The third chapter is devoted to a quite detailed insight into time-dependent scattering theory. The H_3^+ and H_3^- potential energy surfaces (PESs) are presented in the fourth chapter. The fifth chapter shows the results of H^+ and H_2 collisions studied using a non-Born-Oppenheimer method, and the results of H^- and H_2 collisions and its isotopologues which were investigated with quasi-classical trajectories and quantum wave packets.

2 Classical mechanics

The nuclear motion can be solved by classical and quantum mechanics methods. In this chapter we describe the solution of Newton's equations which is a trajectory method. Classical trajectory studies have been widely used in fundamental studies in many areas including the calculation of reaction cross sections, angular distributions, and investigation of reactions as a function of initial and final energy distributions and other observable reaction attributes. In addition, classical trajectories are also applied to get insight into the actual reaction event. Therefore, we can use this method to investigate atomic motions, calculate unobservable opacity functions, and the dependence on features of the potential energy surface.

One of the most useful classical trajectory methods is the quasi-classical version. The term "quasi-classical" is used to denote the manner in which molecules are prepared before collision (i.e., using the correct initial vibrational and rotational quantum numbers as initial conditions). With this the quasi-classical trajectory method assumes that each of the nuclei comprising a chemical system move according to the laws of classical mechanics in the force field derived from the adiabatic electronic energy of the system.

2.1 Specifying the initial parameters

(A) Calculation of internal energies.

From the one-dimensional Schrödinger equation we can determine the discrete eigenvalues and eigenfunctions, and the related vibrational (v) and rotational (j) states for the internal energy $E_{v,j}$. The Schrödinger equation is given as

$$-\frac{\hbar^2}{2\mu}\frac{d^2\Psi_{v,j}(r)}{dr^2} + V_j(r)\Psi_{v,j}(r) = E_{v,j}\Psi_{v,j}(r) \ , \tag{4}$$

where μ is the reduced mass of the system, $V_j(r)$ is the sum of the rotationless potential $V(r)$ and a centrifugal term. The centrifugal potential of a diatomic molecule has the form $[j(j+1) - \Omega^2]\frac{\hbar^2}{2\mu r^2}$, where Ω is the projection of the electronic angular momentum onto the internuclear axis (see Level 8.0 [116], p. 2). We choose the program Level 8.0 [116] to calculate the internal energies. Details will be shown in the result part.

(B) Calculation of turning points.

For the calculation of the vibrational turning points, we use the bisection method ([117], p. 142). The bisection method is a root-finding method which repeatedly bisects an interval and then selects a subinterval in which a root must lie for further processing. The method is applicable when we wish to solve the equation $f(x) = 0$ for the real variable x. For the function $f(x)$, the bisection method needs two starting values of x, which are named as x_1 and x_2, and the root x_r from the function should satisfy $x_1 < x_r < x_2$.

The algorithm goes as follows:

1) the half-interval of x_1 and x_2 was firstly calculated and named as x_3 (i.e., $x_3 = \frac{x_1 + x_2}{2}$). At the same time, the function was evaluated among these three points.

9

2) If the sign of $f(x_3)$ and $f(x_1)$ is different, then the new root lies between x_1 and x_3. Otherwise, the root lies between x_2 and x_3.

3) If the root lies between x_1 and x_3, this interval is bisected, $x_4 = \frac{x_1 + x_3}{2}$. $f(x_1)$, $f(x_3)$, and $f(x_4)$ should be evaluated. Similar calculations should be done for the cases that the root lies between x_2 and x_3.

4) In the next step is one has to check for following root.

5) This procedure is continued until convergence is gained

$$\frac{x_i + x_j}{x_i} \leq \epsilon .\qquad (5)$$

In Eq. (5) ϵ is an arbitrary convergence criterion.

(C) Calculation of the vibrational half-period.

In this part we follow the strategy of Truhlar and Muckerman [90] to calculate the diatomic vibrational half-period $\frac{1}{2}\tau_{BC}^{v,j}$ (see [90], p. 513)

$$\frac{1}{2}\tau_{BC}^{v,j} = (\frac{\mu_{BC}}{2})^{\frac{1}{2}} \int_{r_-}^{r_+} [\varepsilon_{v,j} - V_{BC}(r) - \frac{j(j+1)\hbar^2}{2\mu_{BC}r^2}]^{-\frac{1}{2}} dr .\qquad (6)$$

In Eq. (6) $\varepsilon_{v,j}$ is the internal energy of the BC molecule with quantum values for vibrational state v and rotational state j. μ_{BC} is reduced mass of atom B and C. $r_\pm$ are outer and inner turning points, respectively. r is the internuclear distance between atom B and C. $\hbar$ is a convenient modification of Planck's constant ($\hbar = \frac{h}{2\pi} = 1.05457 \times 10^{-34}$ Js). $V_{BC}(r)$ is the potential of the diatomic when the third atom is far away

$$V_{BC}(r) \equiv V_{BC}(R_2) = \lim_{R_1,R_3 \to \infty} V(R_1, R_2, R_3) .\qquad (7)$$

2.2 The Monte Carlo method

The Monte Carlo method is a class of computational algorithms that rely on repeated random sampling to compute the desired property. Details for this part are given in the book of Wong [118], p. 383–406. Monte Carlo techniques are useful in solving a variety of problems in physics in which one generates random numbers. The Monte Carlo method has the following features:

1) The generator is fast and simple to use.

2) It has the desired statistical properties.

3) A long repeated period is needed.

The repeated period has to be examined before the calculation of trajectories. Random values x between 0 and 1 are generated. With these random numbers we calculate the initial values of coordinates and momenta of the atoms, and the initial reaction impact parameter b.

2.3 The equations of motion

For the collision system A + BC we use the potential energy function as an analytic function which depends on three internuclear distances, i.e., $V \equiv V(R_1, R_2, R_3)$. R_1, R_2 and R_3 are the AB, BC, and AC distances, respectively.

At first, the coordinates of the three atoms system should be defined. The best way is to choose a space-fixed cartesian system consisting of the nine coordinates $x \equiv \{x_i; \ i = 1, ..., 9\}$ of the nuclei A, B, and C, respectively, and the nine momenta $p_x \equiv \{p_{x_i}; \ i = 1, ..., 9\}$ conjugate to these coordinates. The Hamiltonian in this reference coordinate system is given as

$$H(x, p_x) = T(p_x) + V[R_1(x), R_2(x), R_3(x)] \tag{8}$$

with

$$T(p_x) = \sum_{i=1}^{3} \left(\frac{1}{2m_A} p_{x_i}^2 + \frac{1}{2m_B} p_{x_{i+3}}^2 + \frac{1}{2m_C} p_{x_{i+6}}^2 \right) . \tag{9}$$

Hamilton's equations of motion for all nine degrees of freedom are given as

$$\dot{x}_i \equiv \frac{dx_i}{dt} = \frac{\partial H}{\partial p_{x_i}} = \frac{\partial T}{\partial p_{x_i}} \quad (i = 1, ..., 9) , \tag{10}$$

$$\dot{p}_{x_i} \equiv \frac{dp_{x_i}}{dt} = -\frac{\partial H}{\partial x_i} = -\frac{\partial V}{\partial x_i} = -\sum_{k=1}^{3} \frac{\partial V}{\partial R_k} \frac{\partial R_k}{\partial x_i} \quad (i = 1, ..., 9) . \tag{11}$$

The relationship between the three internuclear distances and the reference coordinates is given by the following relation:

$$R_1 \equiv R_{AB} = [\sum_{i=1}^{3}(x_i - x_{i+3})^2]^{\frac{1}{2}}, \quad R_2 \equiv R_{BC} = [\sum_{i=1}^{3}(x_{i+3} - x_{i+6})^2]^{\frac{1}{2}}, \quad R_3 \equiv R_{AC} = [\sum_{i=1}^{3}(x_i - x_{i+6})^2]^{\frac{1}{2}} . \tag{12}$$

Eq. (12) can be used to derive all the necessary terms for $\frac{\partial R_k}{\partial x_i}$. The generalized coordinates that are used for Hamilton's equations are defined as

$$q_i = x_{i+6} - x_{i+3} \quad (i = 1, 2, 3) , \tag{13}$$

$$Q_i = x_i - \frac{1}{(m_B + m_C)}[m_B x_{i+3} + m_C x_{i+6}] \quad (i = 1, 2, 3) , \tag{14}$$

$$S_i = \frac{1}{M}[m_A q_i + m_B q_{i+3} + m_C q_{i+6}] \quad (i = 1, 2, 3) . \tag{15}$$

The total mass M is given as $M \equiv m_A + m_B + m_C$. q_i are the internal coordinates of the diatomic molecule BC. Q_i are the relative coordinates A to BC. S_i defines the center-of-mass coordinate.

Using the generalized coordinates $[q_i, p_i, Q_i, P_i, S_i, P_{S_i}, i = 1, 2, 3]$ the final Hamiltonian can be obtained. p_i, P_i, and P_{S_i} are the conjugate momenta related to p_i, Q_i, and S_i, respectively. The new form of the Hamiltonian is given as

$$H(q, Q, p, P, P_s) = T(q, P, P_s) + V[R_1(q, Q), R_2(q, Q), R_3(q, Q)] , \tag{16}$$

$$T(p, P, P_s) = \sum_{i=1}^{3} \left(\frac{1}{2\mu_{BC}} p_i^2 + \frac{1}{2\mu_{A,BC}} P_i^2 + \frac{1}{2M} P_{S_i}^2 \right) , \tag{17}$$

$$\mu_{BC} \equiv \frac{m_B m_C}{(m_B + m_C)} \,, \qquad \mu_{A,BC} \equiv \frac{m_A(m_B + m_C)}{M} \,. \tag{18}$$

μ_{BC} and $\mu_{A,BC}$ are the reduced masses corresponding to internal and relative motion. The conjugate internuclear distances are

$$R_1 = [\sum_{i=1}^{3} (\frac{m_C}{m_B + m_C} q_i + Q_i)^2]^{\frac{1}{2}}, \qquad R_2 = [\sum_{i=1}^{3} Q_i^2]^{\frac{1}{2}}, \qquad R_3 = [\sum_{i=1}^{3} \frac{m_B}{m_B + m_C} q_i - Q_i)^2]^{\frac{1}{2}} \,. \tag{19}$$

The new internuclear distances R_1, R_2, and R_3 are independent of S_1, S_2, and S_3. P_{s_1}, P_{s_2}, and P_{s_3} are constants of the motion. Hence the term involving P_{s_1}, P_{s_2}, and P_{s_3} can be eliminated from the Hamiltonian. So we get the following twelve equations:

$$\dot{q}_i = \frac{\partial H}{\partial p_i} = \frac{\partial T}{\partial p_i} \qquad (i = 1, 2, 3) \,, \tag{20}$$

$$\dot{Q}_i = \frac{\partial H}{\partial P_i} = \frac{\partial T}{\partial P_i} \qquad (i = 1, 2, 3) \,, \tag{21}$$

$$\dot{p}_i = -\frac{\partial H}{\partial q_i} = -\frac{\partial V}{\partial q_i} = -\sum_{k=1}^{3} \frac{\partial V}{\partial R_k} \frac{\partial R_k}{\partial q_i} \qquad (i = 1, 2, 3) \,, \tag{22}$$

$$\dot{P}_i = -\frac{\partial H}{\partial Q_i} = -\frac{\partial V}{\partial Q_i} = -\sum_{k=1}^{3} \frac{\partial V}{\partial R_k} \frac{\partial R_k}{\partial Q_i} \qquad (i = 1, 2, 3) \,, \tag{23}$$

where $R_i(q, Q)$, $i = 1, 2, 3$ are given by Eq. (19). Details for this section are discussed by Truhlar and Muckerman in Ref [90], p. 508–515.

2.4 Initial conditions

Before integrating Hamilton's equations of motion, the initial values of the coordinates and momenta, i.e., $\{q_i^0, Q_i^0, p_i^0, P_i^0; i = 1, 2, 3\}$, should be specified. For the reaction system of atom A colliding with molecule BC(v, j), where (v, j) are the selected vibrational and rotational states of the molecule BC, the atom A and the center of mass of the BC molecule are defined to lie initially in the yz plane on the $-z$ axis, and the direction of the initial relative velocity vector $\vec{v}_{rel}$ is defined to lie along the $+z$ axis.

Four collision parameters (b, θ, ϕ, η) are obtained by using the Monte Carlo method. b is the impact parameter, θ is the initial azimuthal orientation angle of the BC internuclear axis, ϕ is the initial polar orientation angle of the BC interuclear axis, and η is the initial orientation of the BC angular momentum.

The initial phase angle ξ ($\xi = 0$: the inner turning point; $\xi = \pi$: the outer turning point) and the initial separation between A and the center of mass of BC (ρ) should be specified before the trajectory calculation starts.

The initial values are given as follows:

$$Q_1^0 = 0, \qquad Q_2^0 = b, \qquad Q_3^0 = -(\rho^2 - b^2)^{\frac{1}{2}} \,, \tag{24}$$

$$P_1^0 = 0, \qquad P_2^0 = 0, \qquad P_3^0 = (2\mu_{A,BC}E_{rel})^{\frac{1}{2}} = P_0 \ . \tag{25}$$

Let $r \equiv |q|$ then:

$$q_1^0 = r^0 \sin\theta \cos\phi \ , \qquad q_2^0 = r^0 \sin\theta \sin\phi \ , \tag{26}$$

$$q_3^0 = r^0 \cos\theta \ , \qquad \rho = \rho_0 + \frac{\xi}{2\pi} v_{rel}\tau_{BC}^{v,j} \ . \tag{27}$$

r^0 is initial internuclear distance. With $v = \frac{p}{m}$, Eq. (27), we get for ρ

$$\rho = \rho_0 + \frac{\xi}{2\pi} \frac{P^0\tau_{BC}^{v,j}}{\mu_{A,BC}} \ . \tag{28}$$

$\frac{1}{2}\tau_{BC}^{v,j}$ is the half-period time, which is obtained from Eq. 6. The initial components of the internal (BC) momentum $\vec{p}$ are

$$p_1^0 = J_r \frac{(\sin\phi\cos\eta - \cos\theta\cos\phi\sin\eta)}{r_\pm} \ , \tag{29}$$

$$p_2^0 = -J_r \frac{(\cos\phi\cos\eta + \cos\theta\sin\phi\sin\eta)}{r_\pm} \ , \tag{30}$$

$$p_3^0 = J_r \frac{(\sin\theta\sin\eta)}{r_\pm} \ . \tag{31}$$

$r_\pm$ is the inner or outer turning point, and J_r is the angular momentum of the BC molecule. Details for this section are given in the book of Truhlar and Muckerman [90], p. 511–515.

2.5   The calculation of a single trajectory

After specifying the equations of motion and the initial conditions, the next step is to perform one trajectory. This is accomplished by numerical integration of the equations of motion using the given initial values of $\{q_i, \ Q_i, \ p_i, \ P_i; \ i = 1, \ 2, \ 3\}$. The most popular integrator for trajectory studies of chemical reactions is the fourth-order Runge-Kutta method (see [117], p. 351, and [118], p. 499–502), where for the differential equation $f(y,t) = \frac{dy}{dt}$ the following approximation is used

$$y(t_{k+1}) \approx y(t_k) + \frac{1}{6}h(p + 2q + 2r + s) \tag{32}$$

with

$$p = f(y(t_k), t_k) \ , \qquad\qquad q = f(y(t_k) + \frac{h}{2}p, t_k + \frac{h}{2}) \ ,$$

$$r = f(y(t_k) + \frac{h}{2}q, t_k + \frac{h}{2}) \ , \qquad s = f(y(t_k) + hr, t_k + h) \ . \tag{33}$$

In Eq. (33) t_k is the given time, $y(t_k)$ specifies the given initial values of $\{q_i, \ Q_i\}$, h is a fixed time step.

2.6   Product analysis

Once the trajectory is completed and the product diatomic molecule is tentatively identified, a
new set of generalized coordinates and momenta, in which the Hamiltonian is asymptotically
separable, has to be analyzed:

$$H(\vec{q}, \vec{Q'}, \vec{p}, \vec{P'}) \sim T_{rel}(\vec{P'}) + H_{int}(\vec{q'}, \vec{p'}) \ . \tag{34}$$

$\vec{q'}$ and $\vec{p'}$ are the coordinates and conjugate momenta for the product diatomic molecule and
$\vec{Q'}$ and $\vec{P'}$ are those for the relative motion of the products. The reactant coordinates can be
used to derive the product coordinates $\{q'_i, Q'_i, p'_i, P'_i; \ i = 1, 2, 3\}$ through the following two
steps:

(i) The transformation matrices T and T' define the transformation from reactant to product
coordinates

$$\begin{bmatrix} \vec{q'} \\ \vec{Q'} \end{bmatrix} = \mathbf{T'}\vec{x} \ , \qquad \begin{bmatrix} \vec{q} \\ \vec{Q} \\ \vec{S} \end{bmatrix} = \mathbf{T}\vec{x} \ . \tag{35}$$

(ii) The inverse of the transformation has to be calculated

$$\vec{x} = \mathbf{T}^{-1} \begin{bmatrix} \vec{q} \\ \vec{Q} \\ \vec{S} \end{bmatrix} \ , \tag{36}$$

which leads to

$$\begin{bmatrix} \vec{q'} \\ \vec{Q'} \end{bmatrix} = \mathbf{T'T}^{-1} \begin{bmatrix} \vec{q} \\ \vec{Q} \\ \vec{S} \end{bmatrix} \ . \tag{37}$$

The new conjugate generalized momenta are obtained from

$$p'_i = m\dot{q}'_i \qquad P'_i = \mu \dot{Q}'_i \qquad (i = 1, 2, 3) \tag{38}$$

where

$$\begin{bmatrix} \dot{\vec{q'}} \\ \dot{\vec{Q'}} \end{bmatrix} = \mathbf{T'T}^{-1} \begin{bmatrix} \dot{\vec{q}} \\ \dot{\vec{Q}} \\ \dot{\vec{S}} \end{bmatrix} \ . \tag{39}$$

and

$$\dot{q}_i = \frac{1}{\mu_{BC}} p_i \qquad \dot{Q}_i = \frac{1}{\mu_{A,BC}} P_i \qquad \dot{S}_i = \frac{1}{M} P_{S_i} \qquad (i = 1, 2, 3) \ . \tag{40}$$

For the products AB + C m is μ_{AB} and μ is $\mu_{C,AB}$. For the products AC + B m is μ_{AC} and μ is $\mu_{B,AC}$.
The detailed form of the product coordinates is listed below

AB + C:

$$q_i' = -\frac{m_C}{m_B + m_C}q_i - Q_i \qquad (i = 1, 2, 3), \tag{41}$$

$$Q_i' = -\frac{m_B M}{(m_A + m_B)(m_B + m_C)}q_i + \frac{m_A}{m_A + m_B}Q_i \qquad (i = 1, 2, 3) \ . \tag{42}$$

AC + B:

$$q_i' = \frac{m_B}{m_B + m_C}q_i - Q_i \qquad (i = 1, 2, 3), \tag{43}$$

$$Q_i' = \frac{m_C M}{(m_A + m_C)(m_B + m_C)}q_i + \frac{m_A}{m_A + m_C}Q_i \qquad (i = 1, 2, 3) \ . \tag{44}$$

The Hamiltonian in product coordinates and the internal energies are given as

$$H' = \frac{1}{2\mu}\sum_{i=1}^{3}P_i'^2 + \frac{1}{2m}\sum_{i=1}^{3}p_i'^2 + V(R(\vec{Q'})) \ , \tag{45}$$

$$\tilde{\varepsilon}_{int}' = \frac{1}{2m}\sum_{i=1}^{3}p_i'^2 + V_D(r) \ , \qquad r \equiv [q' \cdot q']^{\frac{1}{2}} \ . \tag{46}$$

The different angular momenta, internal angular momentum $\vec{J_r'}$, relative angular momentum $\vec{J_{rel}'}$, and total angular momentum $\vec{J_{tot}'}$ are listed below

$$\vec{J_r'} = \vec{q'} \times \vec{p'} = (q_2'p_3' - q_3'p_2')\hat{e}_x + (q_3'p_1' - q_1'p_3')\hat{e}_y + (q_1'p_2' - q_2'p_1')\hat{e}_z \tag{47}$$

$$\vec{J_{rel}'} = \vec{Q'} \times \vec{P'} \tag{48}$$

$$\vec{J_{tot}'} = \vec{J_r'} + \vec{J_{rel}'} = \vec{q'} \times \vec{p'} + \vec{Q'} \times \vec{P'} \ . \tag{49}$$

The relative velocity $\vec{v_{rel}'}$ and relative speed is given as

$$\vec{v_{rel}'} = \frac{1}{\mu}\vec{P'}, \qquad \nu_{rel}' = \frac{1}{\mu}(\vec{P'} \cdot \vec{P'})^{\frac{1}{2}} = \frac{1}{\mu}(\sum_{i=1}^{3}P_i'^2)^{\frac{1}{2}} \ . \tag{50}$$

The scattering angle is defined as

$$\theta = cos^{-1}\frac{v_{rel} \cdot v_{rel}'}{\nu_{rel}\nu_{rel}'}, \qquad v_{rel} = \frac{P^0}{\mu_{A,BC}}\hat{e}_z \ . \tag{51}$$

v_{rel} is the initial relative velocity. The diatomic vibrational and rotational energies of the products are given as

$$\tilde{\varepsilon}_{rot}' = min\{V_D(r) + \frac{\vec{J_r'} \cdot \vec{J_r'}}{2mr^2}\} - V_D(r_e), \qquad \tilde{\varepsilon}_{vib}' = \tilde{\varepsilon}_{int}' - \tilde{\varepsilon}_{rot}' \ . \tag{52}$$

This leads to the diatomic rotational "quantum number" j' and diatomic vibrational "quantum number" v' of the products

$$\tilde{j}' = -\frac{1}{2} + \frac{1}{2}[1 + \frac{4\vec{J_r'} \cdot \vec{J_r'}}{\hbar^2}]^{\frac{1}{2}} \tag{53}$$

$$\tilde{v}' = -\frac{1}{2} + \frac{1}{\pi\hbar}\int_{r_-}^{r_+}\{2m[\varepsilon_{int}' - V_D(r) - \frac{J_r' \cdot J_r'}{2mr^2}]\}^{\frac{1}{2}}dr \ . \tag{54}$$

Details for this section are given in Truhlar and Muckerman in Ref. [90], p. 530–534.

2.7 Calculation of the reaction probability

After computing a series of trajectories N_{tot}, the reactive trajectory numbers N_r can be obtained. The reaction probability P_r for the selected initial states (v, j), the initial collision energy E_{rel} and the total number of trajectories N_{tot} is defined by

$$P_r(E_{rel}, v, j) = \frac{N_r}{N_{tot}} . \tag{55}$$

2.8 Reaction cross section

The reaction cross section is defined as an effective area in a plane perpendicular to the initial velocity $\vec{v}$, such that the relative separation vector $\vec{r}$ has to be within that area for a collision to take place. In terms of the opacity function $P(b)$ $(0 \leq P(b) \leq 1)$, which is defined as the fraction of collisions with impact parameter b that lead to a reaction, the reaction cross section σ_R is defined as

$$\sigma_R = \int_0^\infty 2\pi b P(b) db . \tag{56}$$

To account for steric requirements, the reaction probability for a realistic function can be modified by the introduction of a steric factor $p(< 1)$, such that

$$P(b) = \{ {p, \quad b \leq b_{max} \atop 0, \quad b > b_{max}} . \tag{57}$$

b_{max} is the maximum impact parameter b for which a reaction can occur. Substituting Eq. (57) into Eq. (56), the following can be obtained:

$$\sigma_R = \int_0^{b_{max}} 2\pi b P db$$
$$= 2\pi P \int_0^{b_{max}} b db$$
$$= \pi P b_{max}^2 . \tag{58}$$

From Eq. (58) we can conclude that the reaction cross section σ_R depends only on the value of the product $P * b_{max}^2$. In our calculation the maximum impact parameter b_{max} was identified at first, then the reaction probability was determined. In this way the reaction cross section should be reasonably accurate.

2.9 Nonadiabatic multi-surface classical mechanics

In this section the theory of nonadiabatic transitions will be discussed. The Hamiltonian for a molecular system may be written as

$$H(\mathbf{R}, \mathbf{r}) = T_{\mathbf{R}} + H_e(\mathbf{R}, \mathbf{r}), \tag{59}$$

where R and r are the vectors of the nuclear and electronic coordinates, respectively. T_R is the nuclear kinetic energy operator, and H_e is the electronic Hamiltonian, which contains the electronic kinetic energy operator and all Coulomb interactions.

Some chemical systems may be modeled adequately within the framework of the Born-Oppenheimer (BO) approximation. In these systems the nuclear motion is governed by a single BO potential energy surface V, which is the ground-state electronic potential energy surface. The ground-state nuclear wave function ψ_0 may be described with

$$[T_{\mathbf{R}} + V(\mathbf{R}) - E]\psi_0(\mathbf{R}) = 0 , \qquad V(\mathbf{R}) = \langle\phi_0|H_e(\mathbf{R},\mathbf{r})|\phi_0\rangle_{\mathbf{r}} . \tag{60}$$

E is the total energy, ϕ_0 is the ground state electronic wave function, and the integration in Eq. (60) is for the electronic coordinates r.

For non-BO processes, the single ground state surface treatment, described by Eq. (60), is not correct. A new theoretical framework may be developed in terms of a basis set of electronic wave functions ϕ_i, where i labels the electronic states, and optionally one may choose this basis so that ϕ_0 has the same meaning as above. The potential energy surfaces of each electronic state may be defined as

$$V_{ii}(\mathbf{R}) = \langle\phi_i|H_e(\mathbf{R},\mathbf{r})|\phi_i\rangle_{\mathbf{r}}, \tag{61}$$

as well as non-zero off-diagonal matrix elements of the electronic Hamiltonian

$$V_{ij}(\mathbf{R}) = \langle\phi_i|H_e(\mathbf{R},\mathbf{r})|\phi_j\rangle_{\mathbf{r}}. \tag{62}$$

By expanding the multi-state wave function ψ in terms of the electronic basis function

$$\psi(\mathbf{R},\mathbf{r}) = \sum_i \phi_i(\mathbf{R},\mathbf{r})\psi_i(\mathbf{R}), \tag{63}$$

the following equation can be obtained

$$[T_{\mathbf{R}} + V_{ii}(\mathbf{R}) + T_{ii}^{(2)} - E]\psi_i(\mathbf{R}) = -\sum_{j\neq i}[T_{ij}^{(1)}(\mathbf{R}) + T_{ij}^{(2)}(\mathbf{R}) + V_{ij}(\mathbf{R})]\psi_j(\mathbf{R}), \tag{64}$$

with

$$T_{ij}^{(1)} = \frac{-\hbar^2}{2\mu}\langle\phi_i|\nabla_{\mathbf{R}}|\phi_j\rangle \cdot \nabla_{\mathbf{R}} = \frac{-\hbar^2}{2\mu}d_{ij} \cdot \nabla_{\mathbf{R}}, \tag{65}$$

$$T_{ij}^{(2)} = \frac{-\hbar^2}{2\mu}\langle\phi_i|\nabla_{\mathbf{R}}^2|\phi_j\rangle . \tag{66}$$

μ is the reduced mass for the nuclear system, and ∇_R is the nuclear gradient operator, d_{ij} represents the nonadiabatic coupling vectors. From Eq. (64) one can derive that the nuclear motion in each electronic state is controlled by the potential energy surface associated with that state as well as the various coupling terms in Eqs. (62)–(66).

For a given classical path, the electronic motion may be readily obtained by solving the solution to the time-dependent electronic Schrödinger equation [140]. For a two-state system, the time dependence of the electronic population P_1 of the ground state is given by [141, 142]

$$\dot{P}_1 = -2\mathbf{R}_e(a_{12}^* \vec{v} \cdot \vec{d}_{12}), \quad a_{12} = c_1 c_2^* \,, \tag{67}$$

where a_{12} is the electronic coherence of the states 1 and 2. c_i are the complex-value expansion coefficients, and $\vec{v}$ is the nuclear velocity of the trajectory. In the trajectory surface hopping simulations trajectories are propagated in a single electronic state, and the single-surface propagation is interrupted by sudden surface switches or hops to other surfaces. Monitoring the quantum mechanical populations P_i one can determine the location of the surface switches according to the "fewest switches" prescription of Tully [120]. Details for this section are given in Ref. [120, 98].

3 Quantum mechanics

Using quantum mechanics it is impossible to specify simultaneously, with arbitrary precision, both the momentum and the position of a particle [121]. This is different to classical mechanics. Reactive scattering can be investigated time-dependently or time-independently. Applications of time dependent approaches to quantum reactive scattering are becoming increasingly popular [72]. In the present work the time-dependent wave-packet approach will be used.

3.1 The wavepacket propagation

The time-dependent Schrödinger equation will be used to describe the motion of the nuclei in the reaction A + BC

$$i\hbar\frac{\partial}{\partial t}\psi(\vec{R},t) = \hat{H}\psi(\vec{R},t), \tag{68}$$

where $\vec{R}$ is the ensemble of coordinates that define the position of the nuclei. Assuming $\hat{H}$ being time-independent, then the solution of Eq. (68) will be given formally as

$$\psi(\vec{R},t) = \phi(\vec{R})\phi(t) = \exp\left(-\frac{i\hat{H}}{\hbar}t\right)\phi(\vec{R}) . \tag{69}$$

According to Eq. (69) the forward propagation of the wavefunction $\psi(\vec{R},t)$ by time τ is

$$\psi(\vec{R},t+\tau) = \exp(\frac{-i\hat{H}}{\hbar}(t+\tau))\phi(\vec{R}) = \exp(\frac{-i\hat{H}\tau}{\hbar})\psi(\vec{R},t) . \tag{70}$$

The first part of the right hand side in Eq. (70) can be expanded by a Taylor series. Kosloff [122] proposed a global propagator. The main idea is to use a polynomial expansion of the evolution operator:

$$\exp(\frac{-i\hat{H}\tau}{\hbar}) \approx \sum_{n=0}^{N} a_n P_n(-\frac{i\hat{H}}{\hbar}\tau) . \tag{71}$$

Within the Chebychev scheme one approaches this problem in analogy to the approximation of a scalar function. Consider a scalar function $F(x)$ in the interval $[-1,1]$. In this case it is known that the Chebychev polynomial approximation is optimal since the maximum error in the approximation is minimal compared to most of all possible polynomial approximations. In the present approximation of the evolution operator a complex Chebychev polynomial $P_n(\hat{X})$ is used, replacing the scalar function by a function of an operator. In making this change one has to examine the domain of the operator and to adjust it to the range of the definition of the Chebychev polynomial. The range of the definition of these polynomials is from $-i$ to i ($i = \sqrt{-1}$). This means that the Hamiltonian operator has to be renormalized by dividing by $\triangle E$

$$\triangle E = E_{max} - E_{min}, \qquad E_{max} = V_{max} + K_{max}, \qquad E_{min} = V_{min}, \tag{72}$$

19

$$K_{max} = \sum_i \frac{\pi^2 \hbar^2}{2m_i (\triangle q_i)^2} \quad . \tag{73}$$

m_i and $\triangle q_i$ are the mass and grid spacing of the coordinate i, respectively. The Hamiltonian should be limited, i.e., for $V > V_{max}$ we set $V = V_{max}$. So the shifted Hamiltonian is

$$\hat{H}_{norm} = \frac{\hat{H} - \hat{I}(\frac{\triangle E}{2} + V_{min})}{\frac{\triangle E}{2}} \quad . \tag{74}$$

The wavefunction will be approximated in the following way

$$\exp(\frac{-i\hat{H}\tau}{\hbar})\psi(R,t) = \exp(\frac{-i(\frac{\triangle E}{2} + V_{min})\tau}{\hbar}) \sum_{n=0}^{N} (2 - \delta_{n0}) J_n(\frac{\triangle E\tau}{2\hbar}) P_n(-i\hat{H}_{norm})\psi(R,t) \quad . \tag{75}$$

The complex Chebychev polynomials P_n fulfil a recursion relation

$$P_{n+1} = -2i\hat{H}_{norm}P_n + P_{n-1} \quad . \tag{76}$$

The first term in the right-hand side of Eq. (75) is a phase shift compensating the shift in the energy scale. The first three Chebychev polynomials are given as follows

$$P_0(-ix) = 1 \ , \quad P_1(-ix) = -2ix \ , \quad P_2(-ix) = -4x^2 + 1 \ . \tag{77}$$

The $J_n(\alpha)$ are Bessel functions with

$$(2 - \delta_{n0})J_n(\alpha) = \int_{-i}^{i} \frac{\exp(i\alpha x)P_n(x)dx}{(1 - x^2)^{\frac{1}{2}}}, \quad \alpha = \frac{\triangle E\tau}{\hbar}. \tag{78}$$

The maximum expansion number term N is approximately $N \approx \frac{\triangle E\tau}{2\hbar}$.

3.2 Time evolution

In 1998 Gray and Balint-Kurti [85] developed a new version of the Chebyshev expansion, where only the real part of the wavepacket was used. The backward propagation can be expressed by

$$\psi(R, t - \tau) = \exp(\frac{i\hat{H}\tau}{\hbar})\psi(R,t). \tag{79}$$

Combining Eq. (70) with Eq. (79) leads to

$$\psi(R, t + \tau) = -\psi(R, t - \tau) + 2\cos(\frac{\hat{H}\tau}{\hbar})\psi(R,t). \tag{80}$$

Eq. (80) does not include $i = \sqrt{-1}$, so the real and imaginary parts of ψ can be propagated independently. Using the notations

$$q(R,t) = Re[\psi(R,t)] \ , \quad p(R,t) = Im[\psi(R,t)] \tag{81}$$

we get for Eq. (80)

$$q(R, t + \tau) = -q(R, t - \tau) + 2\cos(\frac{\hat{H}\tau}{\hbar})q(R,t). \tag{82}$$

In the next step the real part of the wavefunction $q(R,t)$, according to the initial condition $\psi(t=0) = q(0) + ip(0)$, is obtained. Eq. (70) can be rearranged $(t=0)$

$$q(R, \tau) = \cos(\frac{\hat{H}\tau}{\hbar})q(R, t=0) + \sin(\frac{\hat{H}\tau}{\hbar})p(R, t=0) \ . \tag{83}$$

Eq. (82) is repeatedly used to obtain $q(t)$ for discrete time steps τ. In the calculation a finite size of coordinates in r and R is used. Without taking any precautions into account the wavefunction would be reflected at the boundaries of the grid. This had to be considered. One approach to minimize the reflection is to periodically absorb the wavefunction in a small region of the grid close to the boundary [123]. Absorption is carried out in both the reactant and product channels at every time step τ, with τ not too large ($\tau < 1$ fs). For Jacobi coordinates the absorption form is

$$\hat{A}_{n,k,n',k'} = \delta_{n,n'}\delta_{k,k'}\hat{A}_R(R_k)\hat{A}_r(r_n) \tag{84}$$

with

$$\hat{A}_R(R_k) = \begin{cases} \exp[-C_{abs}^R(R_k - R_{abs})^2], & R_k > R_{abs} \\ 1, & R_k \leq R_{abs} \end{cases} \tag{85}$$

$$\hat{A}_r(r_n) = \begin{cases} \exp[-C_{abs}^r(r_n - r_{abs})^2], & r_n > r_{abs} \\ 1, & r_n \leq r_{abs}. \end{cases} \tag{86}$$

The absorption can be formally included into Eq. (82)

$$q(R, t + \tau) = \hat{A}[-\hat{A}q(R, t - \tau) + 2\cos(\frac{\hat{H}\tau}{\hbar})q(R, t)]. \tag{87}$$

3.3   The A–BC system

3.3.1   Jacobi coordinates

There exist several coordinate systems to describe triatomic molecules. In the case of scattering Jacobi coordinates are appropriate to describe the arrangements A + BC (reactant coordinates) and C + AB (product coordinates) (Fig. 3). These sets of coordinates can be used simultaneously either during the complete propagation or by transforming the wavepacket from one set of initial coordinates to another set of product coordinates. Miller [124] provided a general expression that includes all the reactant and the product arrangements

$$\psi_{\gamma_1, n_1} = \sum_v \phi_v^a(r_a)f_{av \leftarrow \gamma_1 v_1}(R_a) + \sum_v \phi_v^b(r_b)f_{bv \leftarrow \gamma_1 v_1}(R_b) + \sum_v \phi_v^c(r_c)f_{cv \leftarrow \gamma_1 v_1}(R_c), \tag{88}$$

where $\gamma_1 = a(A + BC)$, $b(B + AC)$, or $c(C + AB)$ labels the different arrangements of the atoms. $\{\phi_v^a\}$, $\{\phi_v^b\}$ and $\{\phi_v^c\}$ are the vibrational eigenfunctions of the diatoms BC, AC, and AB, respectively.

This approach is similar to the linear combination of atomic orbitals (LCAO) for molecular orbitals. For example, the electronic diatomic molecular orbital $\chi(r)$ can be expanded as

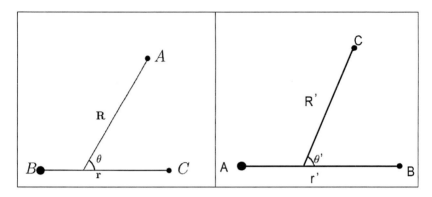

Figure 3: Reactant A–BC and product C–AB Jacobi coordinates systems.

$$\chi(r) = \sum_i a_i \phi_i^a(r_a) + \sum_i b_i \phi_i^b(r_b), \tag{89}$$

where r_a are the coordinates of the electron with respect to nucleus a and r_b corresponding to that of nucleus b. Eq. (88) shows an expansion of the wavepacket for the entrance channel $\gamma_1 = a$ (A + BC) and for the exit channel coordinates b(B + AC) and c(C + AB).

3.3.2   Space-fixed and body-fixed Jacobi coordinates

For a triatomic system there are 9 degrees of freedom. The Jacobi coordinates are used only for the internal degrees of freedom of the system, while the rotational and translational motion (3 + 3 degrees of freedom) are described differently. Therefore, the molecule defined by three Jacobi coordinates should be "inserted" into a three dimensional cartesian coordinate system. In 1974 Pack [125] proposed an improved method. The coordinates are shown in Fig. 4. At first, one has a laboratory-fixed system ("space-fixed coordinates"-SF) and expresses the complete wavefunction by using this system. Second, one defines a body fixed coordinates system BF in such a way that the "Z" axis lies along the R Jacobi coordinate (atom-diatom distance) and has the origin in the centre of mass of the complete triatomic A–BC system. We use $\vec{l}$ to express the relative angular momentum for the motion of atom A relative to BC, and $\vec{j}$ to express the rotational angular momentum of BC. The total angular momentum of the system is $\vec{J}$ ($\vec{J} = \vec{l} + \vec{j}$). If the eigenfunctions of the relative angular momentum $\vec{l}$ are described by spherical harmonics $Y_{l,m_l}(R')$ and the rotational motion is described by the function $Y_{j,m_j}(r')$ then the common set of eigenfunctions for the operators $\hat{J}^2$, $\hat{J}_z$, and $\hat{j}^2$ can be obtained [125, 126] as

$$y_{jl}^{JM}(\hat{r}', \hat{R}') = \sum_{m_j=-j}^{j} \sum_{m_l=-l}^{l} C(jlJ; m_j, m_l, M) Y_{jm_j}(\hat{r}') Y_{lm_l}(\hat{R}') . \tag{90}$$

The complete wavefunction is

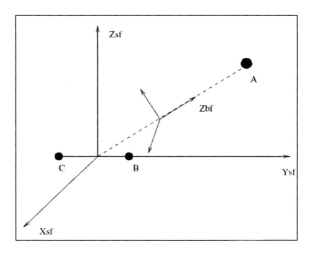

Figure 4: A–BC space-fixed (SF) and body-fixed (BF) Jacobi coordinates systems.

$$\psi^{JMjlv} \equiv \sum_{j''} \sum_{l''} \sum_{v''} R^{-1} G^{Jjlv}_{j'',l'',v''}(R) \chi_{j'',v''}(r) y^{JM}_{j''l''}(\hat{r}', \hat{R}') , \qquad (91)$$

including the radial channel wavefunctions $G^{Jjlv}_{j'',l'',v''}(R)$ and the diatomic vibrational wavefunctions $\chi_{j'',v''}(r)$. Eq. (91) presents the space-fixed formulation of the wavefunction. The Wigner rotation matrix D connects the space-fixed wavefunctions and the body-fixed wavefunctions

$$\psi^{JMjlv} = \sum_{\lambda} D^{J}_{\lambda M}(\omega) \frac{1}{Rr} \phi^{Jj\lambda}_{BF} . \qquad (92)$$

λ is a body-fixed z component of the total angular momentum J and ω is the angle relating space-fixed and body-fixed coordinate systems.

3.4 The Hamiltonian of the 3-atomic system

The action of the Hamiltonian operator on the wavefunction in Jacobi coordinates (R, r, θ) for the body-fixed frame presentation leads to

$$\hat{H}^J \phi^{J\lambda}(R,r,\theta,t) = \{\hbar^2 \{ -\frac{1}{2\mu_R} \frac{\partial^2}{\partial R^2} - \frac{1}{2\mu_r} \frac{\partial^2}{\partial r^2} \}$$

$$-(\frac{1}{2\mu_R R^2} + \frac{1}{2\mu_r r^2})(\frac{1}{\sin\theta} \frac{\partial}{\partial \theta} \sin\theta \frac{\partial}{\partial \theta} - \frac{\lambda^2}{\sin^2\theta})$$

$$+\frac{1}{2\mu_R R^2}(J(J+1) - 2\lambda^2) + V(R,r,\theta)\} \phi^{J\lambda}(R,r,\theta,t)$$

$$+C^J_{\lambda,\lambda-1} \phi^{J\lambda-1}(R,r,\theta,t) + C^J_{\lambda,\lambda+1} \phi^{J\lambda+1}(R,r,\theta,t) . \qquad (93)$$

The last two terms in Eq. (93) present the Coriolis coupling [127], i.e., the coupling of the rotational and vibrational wavefunction with

23

$$C^J_{\lambda,\lambda\pm1} = -\frac{\hbar^2[J(J+1) - \lambda(\lambda\pm1)]^{\frac{1}{2}}[j(j+1) - \lambda(\lambda\pm1)]^{\frac{1}{2}}}{2\mu_R R^2}. \tag{94}$$

The reduced masses μ_R and μ_r are

$$\mu_R = \frac{m_a(m_b + m_c)}{m_a + m_b + m_c} \ , \qquad \mu_r = \frac{m_b m_c}{m_b + m_c} \ . \tag{95}$$

The complete Hamiltonian includes four independent parts. The first part computes the contribution of the kinetic energy, the second part computes the contribution of the rotational energy of the diatomic molecule, the third part computes the potential energy, and the fourth part computes the Coriolis coupling.

3.4.1   The kinetic energy terms

In Jacobi coordinates, along the R and r coordinates, the kinetic energy operator is of the form

$$\hat{T}_{kin} = -\frac{\hbar^2}{2\mu}\frac{\partial^2}{\partial x^2} \ , \qquad (x = r, R, \quad \mu = \mu_r, \mu_R) \ . \tag{96}$$

The most convenient method for dealing with these radial terms is the Fast Fourier transform method (FFT). The FFT calculation for the kinetic energy at the point x is performed in 3 steps. First, the Fourier transform of $\psi(x)$ leads to the momentum representation of the wavefunction $\psi(k)$

$$\psi(k) = \frac{1}{\sqrt{2\pi}}\int_{-\infty}^{\infty}\exp(-ikx)\psi(x)dx = fft\{\psi(x)\} \ . \tag{97}$$

In the next step, $\psi(k)$ has to be multiplied with $\frac{(\hbar k)^2}{2\mu}$. This is a local operation at each grid point for the momentum grid point. In the third step the momentum representation of the wavefunction is transformed back to the space representation by an inverse FFT^{-1}

$$\phi(x) = -\frac{\hbar^2}{2\mu}\frac{\partial^2}{\partial x^2}\psi(x) = fft^{-1}\{\frac{(\hbar k)^2}{2\mu}fft\{\psi(x)\}\} \ . \tag{98}$$

3.4.2   The angular kinetic energy terms

In this section we present the angular part of the Hamiltonian operator (see Eq. (93)). The rotational operator for the diatomic molecule is

$$\hat{T}_{rot} = -\frac{\hbar^2}{2}(\frac{1}{\mu_R R^2} + \frac{1}{\mu_r r^2})[\frac{1}{\sin\theta}\frac{\partial}{\partial\theta}(\sin\theta\frac{\partial}{\partial\theta}) - \frac{j_z^2}{\sin\theta^2}] \ . \tag{99}$$

The centrifugal potential is proportional to $(\frac{j_z^2}{\sin\theta^2})$ and imposes a boundary condition that $\psi(R, r, \theta)$ varies with θ as θ^{j_z} at small θ and as $(\pi - \theta)^{j_z}$ at $\theta \to \pi$. The angular eigenfunctions of the first part of the operator are the Legendre polynomials $P_j(\cos\theta)$

$$[-\frac{1}{\sin\theta}\frac{\partial}{\partial\theta}(\sin\theta\frac{\partial}{\partial\theta})]P_j(\cos\theta) = j(j+1)P_j(\cos\theta) \ . \tag{100}$$

The rotational energy can be presented by the finite basis representation (FBR) as

$$T_j^{FBR} = \frac{1}{2\mu}j(j+1)\delta_{jj'} .\tag{101}$$

According to the discrete variable representation (DVR) method [128, 80, 81] the location of the grid points can be defined by the Gauss-type quadrature rule for the associated Legendre functions $P_l^\Omega(\cos(\theta)_{i,\Omega}) = 0$. The grid points (for $\cos(\theta)_{i,\Omega}$, $i = 1, 2, \cdots, n_\Omega$, $\Omega = 0, 1, \cdots, J$) are chosen as roots of the corresponding associated Legendre polynomials. n_Ω are the number of grid points in channel Ω. Light and coworkers [81] calculated the weights for the associated Legendre quadrature using the Christoffel-Darboux formula

$$\frac{1}{w_{i,\Omega}}\delta_{i,j} = \sum_{l=0}^{n_\Omega} P_l^\Omega(\cos(\theta_i))P_l^\Omega(\cos(\theta_j)) .\tag{102}$$

For every channel the associated Legendre quadrature formula can be expressed by a grid representation

$$\int_{-1}^{1} f(\cos(\theta))d(\cos(\theta)) = \frac{1}{k_\Omega}\sum_i w_{i,\Omega}f(\cos(\theta)_{i,\Omega}) , \qquad k_\Omega = \sum_i w_{i\Omega} .\tag{103}$$

The DVR and FBR representations can be transformed to each other with the matrix U_Ω^{jl}

$$U_\Omega^{jl} = \sqrt{w_{j,\Omega}}P_{l,\Omega}(\cos(\theta_j)) , \qquad j,l = 1, \cdots, n_\Omega .\tag{104}$$

3.4.3 The potential energy

The action of the potential energy operator on the wavefunction is calculated by evaluating the potential at the grid points and multiplying it by the value of the representation of the wavefunction at those points, i.e.,

$$\hat{V}_{DVR}|\psi> = V_{DVR}(r_e, R_m, \cos(\theta_i))\sqrt{w_i}\psi_{DVR}(\cos(\theta_i)) .\tag{105}$$

w_i are the corresponding DVR weights.

3.5 Preparation of the initial wavepacket

Balakrishnan et al. [129] have shown that the initial wavepacket can be expressed as a Gauss-function along the R coordinate multiplied by the initial state ro-vibrational wavefunction of the diatomic molecule

$$\psi(R, r, t = 0) = \frac{1}{\pi^{\frac{1}{4}}\sqrt{\sigma}}\exp(ik_0 R)\exp(-\frac{(R - R_0)^2}{2\sigma^2})\chi(r) .\tag{106}$$

R is the distance between the separated atom and the reduced centre of mass of the diatomic system. r_i is the internal coordinate of the diatom and $\chi(r)$ is the initial wavefunction of the diatomic molecule.

As indicated by Heisenberg's uncertainty principle relation $\Delta x \Delta p \geq \frac{\hbar}{2}$, we can evaluate for Gaussian wavepackets the lowest value $\Delta x \Delta p = \frac{\hbar}{2}$. The initial wavepacket should be chosen not only suitable to the momentum but also to the space part. If the spreading of the momentum is very narrow, then it needs a very large grid to install the complete wavepacket. If the energy is very large, then the distance between two successive points of the grid should be very small. This means that one needs a lot of points to store the wavepacket values.

3.6 Analysis of the propagated wavepacket

The scattering matrix (called S–matrix) $S_{j \leftarrow i}$ relates the initial state and the final state of a physical system undergoing a scattering process. It is defined as the unitary [130] matrix connecting asymptotic particle states in the scattering channels

$$S_{j \leftarrow i} = \langle \psi_{out}^{+} | \psi_{in}^{-} \rangle \ . \tag{107}$$

$|\psi_{in}^{-}\rangle$ is the quantum state of the system at an initial time in the entrance channel, and $\langle \psi_{out}^{+}|$ is the quantum state of the system at a final time in the exit channel. There are several methods for computing the S–matrix and reaction probabilities from complete wavepackets. In all these methods the grid representation is separated into two pieces (see Fig. 5). The first piece is the interaction region, i.e., in these parts the potential derivatives with respect to the distance atom-diatom should be considered. The second piece is the asymptotic region where the potential is nearly constant. In these regions the derivatives of the potential with respect to the reactive coordinates are neglected. In the following section we mainly discuss the flux analysis method.

3.6.1 The flux analysis method

The flux analysis method is widely used in wavepacket calculations [129, 133, 136]. The main idea of this method is to calculate the flux of the wavepacket in the exit channel going through an analysis line $r = r_0$ in the product region (see Fig. 5). The reaction probability is given as

$$P_R(E) = \frac{\hbar}{\mu} Im[\langle \Psi(R, r_0, E) | \frac{\partial \Psi(R, r_0, E)}{\partial r} \rangle] \ , \qquad \Psi(R, r_0, E) = \psi(R, r_0, E)/A_e \ . \tag{108}$$

The wavefunction $\psi(R, r_0, E)$ is energy dependent and can be calculated from the Fourier transform of the time-dependent wavefunction $\psi(R, r, t)$ along the analysis line $r = r_0$. The potential energy in the analysis region is assumed to be constant. This Fourier transform leads to

$$\psi(R, r_0, E) = \frac{1}{\sqrt{2\pi}} \int_{-\infty}^{\infty} \psi(R, r, t) \exp \frac{iEt}{\hbar} dt|_{r=r_0} \ . \tag{109}$$

The A_e factor is given as

$$A_e = (\frac{\mu}{\hbar k_{n0}})^{(\frac{1}{2})} A_{n0}(k_{n0}) \ , \qquad A_{n0}(k_{n0}) = \frac{1}{\sqrt{2\pi}} \int_0^{\infty} \int_0^{\infty} \Psi(R, r, t=0)\phi(r) \exp(-ik_{n0}R) dr dR \ . \tag{110}$$

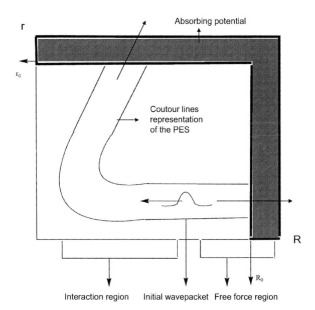

Figure 5: Analysis of the wavepacket propagation in Jacobi coordinates (R, r, θ) for fixed angle θ. (note: R_0 is the position of the analysis line in the reactant channel, r_0 is the position of the analysis line in the product channel).

μ is the reduced mass in the exit channel. k_{n0} is the wavenumber of the plane waves in the entrance channel

$$k_{n0} = \sqrt{\frac{2\mu(E - \epsilon_0)}{\hbar}} \ . \tag{111}$$

E is the total energy of the wavepacket, and ϵ_0 is the initial energy of the diatomic molecule. The final expression for the reaction probability, using the flux formulation, is

$$P_R(E) = \frac{\hbar^2 k_{n0}}{\mu^2 |A_{n0}(K_{n0})|^2} Im[\langle \Psi(R, r_0, E) | \frac{\partial \Psi(R, r_0, E)}{\partial r} \rangle], \quad J = 0 \ . \tag{112}$$

For $J > 0$ this relationship should consider the coupling of the different Ω channels

$$P_R^J(E) = \frac{\hbar^2 k_{n0}}{\mu^2 |A_{n0}(K_{n0})|^2} \sum_{\Omega=0}^{J} Im[\langle \Psi_\Omega(R, r_0, E) | \frac{\partial \Psi_\Omega(R, r_0, E)}{\partial r} \rangle] \ . \tag{113}$$

3.7   The cross section

In the classical trajectory part we have introduced the differential cross section as

$$\sigma_R = \int_0^{b_{max}} 2\pi b P(b) db = \int_0^{b_{max}} 2\pi b P^J db \ , \tag{114}$$

where P^J is the reaction probability corresponding to a given angular momentum J, which is defined as

$$\vec{J} = \vec{r} \times \vec{p} \ . \tag{115}$$

Comparing the classical impact parameter b with the angular momentum J we can write in quantum mechanics formally

$$bk \approx J + \frac{1}{2} \quad or \quad b \approx \frac{1}{k}(J + \frac{1}{2}) \ . \tag{116}$$

k is the wave vector of the projectile $(\vec{p} = \hbar \vec{k})$. As a result we can formulate the cross section based on wavepacket calculations as

$$\sigma^{tot} = \frac{\pi}{k^2} \sum_{J=0}^{\infty} (2J + 1) P^J \ . \tag{117}$$

28

4 H_3^+ and H_3^- PES

4.1   The H_3^+ potential energy surface

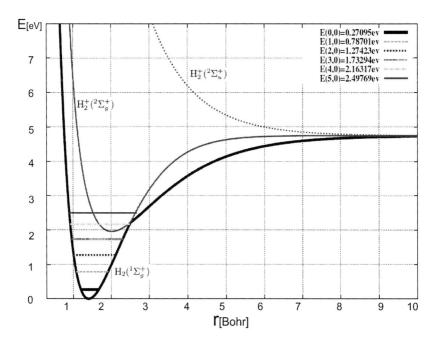

Figure 6: H_3^+: A cut through the three lowest adiabatic PESs at $R(H^+ + H_2) = \infty$. The adiabatic potential curves of H_2 and H_2^+, and the energies $E(v, j)$ of the six lowest vibrational states of H_2 are presented.

There is a large number of theoretical studies which were essential in the identification and understanding of the H_3^+ ion. The study of the electronic potential energy surface (PES) is the first essential ingredient in these models. To the best of my knowledge, there are three kinds of PESs for the H_3^+ ion. The first family is the one including local *ab initio* PESs. Meyer et al. MBB [23] obtained one of these potentials in 1986. They used a full CI calculation at 69 points with the highest energy part 25000 cm^{-1} above the minimum of the PES. Because the PES is a local fit of energy points, the dissociation region is not described highly accurate. The RKJK PES [24] belongs to the same family as the MBB PES. It is a local potential using the same geometric grid points. It relies on a configuration interaction method including explicitly one r_{12} linear term in the wavefunction CI-R12, and the RKJK PES is more accurate than the MBB PES. Later, a PES with even higher accuracy [25, 26] was obtained using correlated

Gaussian functions and where adiabatic and relativistic corrections were included. Jaquet [27] improved this PES in 2002 and improved the PES to regions of higher energy. The semi-empirical fits of PESs are the second family. They are obtained by iterative adjustment of the potential until a good agreement with the experimental spectroscopic data is achieved. The DMT potential [28] is an example of this kind of PES. It is achieved by adjustment to 243 energy levels with a standard deviation of 0.053 cm^{-1}. The third family of PESs presents so called global *ab initio* potential energy surfaces. They dependent on a large number of *ab initio* calculations which cover the whole configuration space. These PESs are useful to calculate highly excited rovibrational states and dynamical processes above the dissociation threshold: PPT [29], PPKT [30], ARTSP [31, 32], VLABP [33], VAV [39] PESs belong to this family PES. The VAV PES which constructs partly a global, diabatic, multivalued PES for singlet H_3^+ is based on a method proposed by Varandas [40].

For the singlet state of the H_3^+ ion there are two PESs available in the literature that can be used to study nonadiabatic processes. The first PES is the one of Preston and Tully (1971) [34]. It is qualitatively correct, but inaccurate even at low energies. The second PES is the KBNN [35] surface, which introduces a better description of the avoided crossing.

For the investigation of the adiabatic and non-adiabatic $H^+ + H_2(v, j)$ reaction, we selected the KBNN [35, 143] potential energy surface. This system has two different electronic channels in the energy range considered:

$$H_2(^1\Sigma_g^+) + H^+ \rightarrow H_2(^1\Sigma_g^+) + H^+, \tag{118}$$

$$\rightarrow H_2^+(^2\Sigma_g^+) + H \quad \text{(charge transfer).} \tag{119}$$

Fig. 6 shows for the asymptotic region at $R(H^+ + H_2) = \infty$ the energy dependence along $r(\text{H–H})$ for the ground and the two lowest excited singlet-states of H_3^+, which represent H_2 and the H_2^+ cation. In addition, vibrational energies for H_2 with initial vibrational states $v = 0$–5 and rotational state $j = 0$ are plotted.

The ground state PES of H_3^+ has a deep minimum of 4.608 eV [33]. It has three asymptotic valleys, each of them corresponding to three different arrangements, which correlate to the $H^+ + H_2(X^1\Sigma_g^+)$ potential. In Fig. 6 it is observed that there is a crossing between the surface of the ground state and the first excited state. At the crossing point the Jacobi diatomic coordinate is $r = 2.50679$ Bohr for the energy $E \approx 2.21$ eV. Near the crossing seam the electron can "jump" between the adiabatic ground state and the first excited state, which we call a non-adiabatic transition.

4.2 H_3^- PES

At present, there are three quite accurate *ab initio* potential energy surfaces (PESs) of H_3^- available: the PES by Stärck and Meyer (SM) [50], by Panda and Sathyamurthy (PS) [53], and by Ayouz et al. (AY) [54]. In addition, Belyaev, Tiukanov and others [55, 56, 57, 58, 59, 60] have obtained PESs of excited electronic states of H_3^- and their non-Born-Oppenheimer couplings with the ground state. The excited electronic states are unstable with respect to electron

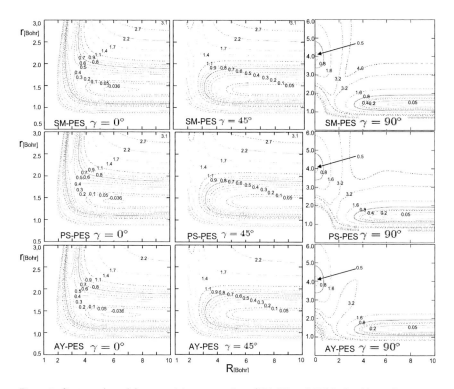

Figure 7: Contour plots of the potential energy surfaces (SM, PS, and AY) in Jacobi coordinates R, r for different angles γ as indicated in each box. The energy unit is eV.

autodetachment.

In 1993 Stärck and Meyer [50] determined an accurate *ab initio* potential energy surface for the reaction $H_2 + H^-$ based on MR-CI and CEPA(2) calculations. They defined modified interatomic distances $\overline{r_i}$ from which modified scattering coordinates $\overline{R}$, $\overline{r}$ and $\overline{\gamma}$ are derived in the usual way. For $r_1 < r_2 < r_3$ the coordinates $\overline{r_i}$ will be given as $\overline{r_3} = r_3$, $\overline{r_1} + \overline{r_2} = r_1 + r_2$, and

$$\overline{r_2} - \overline{r_1} = ((r_2 - r_1)^2 + 4\epsilon^2)^{\frac{1}{2}}, \quad \epsilon = P_1 \exp[-p_2(\frac{r_2 - r_1}{2})^2], \tag{120}$$

where p_1 and p_2 are fitting parameters.

The "Morse" coordinate is given as

$$r' = \frac{1 - \exp[p_3(r_e - \overline{r})]}{p_3} \tag{121}$$

31

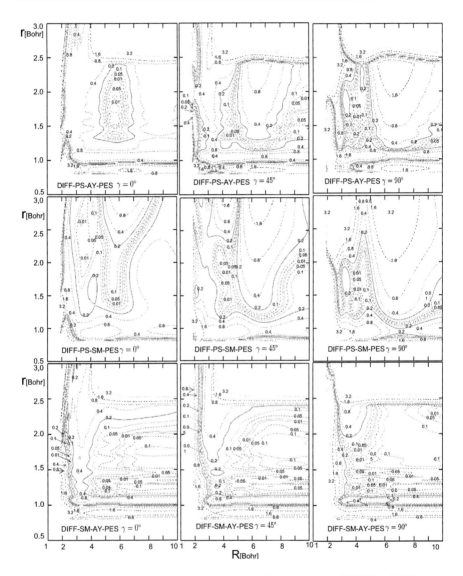

Figure 8: Contour plots of the differences of the potential energy surfaces SM, PS, and AY in Jacobi coordinates R, r for different angles γ values as indicated in each box. The energy unit is eV.

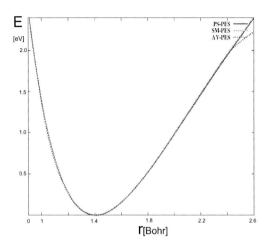

Figure 9: The H_2 potential curves in H_3^- for the SM-PES, PS-PES, and AY-PES.

with $r_e = 1.4$ Bohr. R' is defined as $R' = \overline{R} - 3.0$.
The long-range potential is given as

$$V_{lr} = \sum_{n,l} C_{nl}(\overline{r})\overline{R}^{-n} D_n(\overline{R}, \overline{r}, \overline{\gamma}) P_l(\cos \overline{\gamma}), \tag{122}$$

where the damping function D_n has the form of an incomplete gamma function

$$
\begin{aligned}
D_n(x) &= 0, & x < 0, \\
&= 1 - \sum_{i=0}^{n} \frac{x^i}{i!} \exp(-x), & x \geq 0, \\
x &= \overline{R}[p_4 + p_5 r' + p_6 R' + p_7 P_2(\cos\overline{\gamma}) + p_8(n-3)],
\end{aligned}
$$

with non-linear parameters p_4 to p_8.
The exchange repulsion is cast in an exponential form as

$$V_{ex} = A_{ex} \exp[R'(p_9 + p_{10})P_2(\cos\overline{\gamma})], \tag{123}$$

$$A_{ex} = c_1 + c_2 r' + c_3 P_2(\cos\overline{\gamma}) + c_4 R' + c_5 r'^2 + c_6 P_2(\cos\overline{\gamma})R' + c_7 R'^2 + c_8 r' P_2(\cos\overline{\gamma}). \tag{124}$$

The chemical binding, responsible for the low barrier for linear structures, is assumed to stem from resonance interactions between the σ_u orbital of H_2 and the diffuse s orbital of H^-. This energy is written as

$$V_{rs} = (1 - (1 + 4\beta_{rs}^2)^{\frac{1}{2}})(c_9 + c_{10}\cos^2\overline{\gamma}), $$
$$\beta_{rs} = \exp(p_{11}R' + p_{12}r')p_{13}\cos\overline{\gamma}. \tag{125}$$

The total potential energy is given as

$$V = V_{lr} + V_{ex} + V_{rs} + V_{as}, \tag{126}$$

where V_{as} represents the asymptotic energy.

We continued this work to get the derivatives of the potential energy with respect to the three internuclear distances (r_1, r_2, r_3)

$$\frac{\partial V}{\partial r_i} = \frac{\partial V_{lr}}{\partial r_i} + \frac{\partial V_{ex}}{\partial r_i} + \frac{\partial V_{rs}}{\partial r_i} + \frac{\partial V_{as}}{\partial r_i}, \qquad i = 1, 2, 3, \tag{127}$$

with $\frac{\partial V_{as}}{\partial r_i} = 0$.

According to the chain rule the following equations can be obtained:

$$\frac{\partial V_{lr}}{\partial r_i} = \frac{\partial V_{lr}}{\partial \overline{R}} \frac{\partial \overline{R}}{\partial r_i} + \frac{\partial V_{lr}}{\partial \overline{r}} \frac{\partial \overline{r}}{\partial r_i} + \frac{\partial V_{lr}}{\partial \cos \overline{\gamma}} \frac{\partial \cos \overline{\gamma}}{\partial r_i}, \qquad i = 1, 2, 3, \tag{128}$$

$$\frac{\partial V_{ex}}{\partial r_i} = \frac{\partial V_{ex}}{\partial \overline{R}} \frac{\partial \overline{R}}{\partial r_i} + \frac{\partial V_{ex}}{\partial \overline{r}} \frac{\partial \overline{r}}{\partial r_i} + \frac{\partial V_{ex}}{\partial \cos \overline{\gamma}} \frac{\partial \cos \overline{\gamma}}{\partial r_i}, \qquad i = 1, 2, 3, \tag{129}$$

$$\frac{\partial V_{rs}}{\partial r_i} = \frac{\partial V_{rs}}{\partial \overline{R}} \frac{\partial \overline{R}}{\partial r_i} + \frac{\partial V_{rs}}{\partial \overline{r}} \frac{\partial \overline{r}}{\partial r_i} + \frac{\partial V_{rs}}{\partial \cos \overline{\gamma}} \frac{\partial \cos \overline{\gamma}}{\partial r_i}, \qquad i = 1, 2, 3, \tag{130}$$

with

$$\frac{\partial V_{lr}}{\partial \overline{R}} = \frac{\partial V_{lr}}{\partial R'} \frac{\partial R'}{\partial \overline{R}}, \qquad \frac{\partial V_{ex}}{\partial \overline{R}} = \frac{\partial V_{ex}}{\partial R'} \frac{\partial R'}{\partial \overline{R}}, \qquad \frac{\partial V_{rs}}{\partial \overline{R}} = \frac{\partial V_{rs}}{\partial R'} \frac{\partial R'}{\partial \overline{R}}, \tag{131}$$

$$\frac{\partial V_{lr}}{\partial \overline{r}} = \frac{\partial V_{lr}}{\partial r'} \frac{\partial r'}{\partial \overline{r}}, \qquad \frac{\partial V_{ex}}{\partial \overline{r}} = \frac{\partial V_{ex}}{\partial r'} \frac{\partial r'}{\partial \overline{r}}, \qquad \frac{\partial V_{rs}}{\partial \overline{r}} = \frac{\partial V_{rs}}{\partial r'} \frac{\partial r'}{\partial \overline{r}}, \tag{132}$$

$$\frac{\partial V_{lr}}{\partial \cos \overline{\gamma}} = \frac{\partial V_{lr}}{\partial P_2(\cos \overline{\gamma})} \frac{\partial P_2(\cos \overline{\gamma})}{\partial \cos \overline{\gamma}}, \qquad \frac{\partial V_{ex}}{\partial \cos \overline{\gamma}} = \frac{\partial V_{ex}}{\partial P_2(\cos \overline{\gamma})} \frac{\partial P_2(\cos \overline{\gamma})}{\partial \cos \overline{\gamma}}. \tag{133}$$

$P_2(\cos \overline{\gamma})$ is a Legendre polynomial. One should take into account the condition $r_1 < r_2 < r_3$ when calculating theses derivatives.

Panda and Sathyamurthy [53] constructed a global analytical potential energy surface (PES) for the ground state of H_3^- in 2004. Their PES was generated using the suite of programs MOLPRO [137] for a grid of energy points using the center-of-mass separation R ranging from 2 to 13 Bohr and intramolecular bond distances r ranging from 1.0 to 4.0 Bohr for the Jacobi angle $\gamma = 0, 30, 60, 90°$. The potential energy function for the triatomic system ABC will be expanded as

$$V_{ABC}(R_1, R_2, R_3) = V_A^{(1)} + V_B^{(1)} + V_C^{(1)} + V_{AB}^{(2)}(R_1) + V_{BC}^{(2)}(R_2) + V_{AC}^{(2)}(R_3) + V_{ABC}^{(3)}(R_1, R_2, R_3). \tag{134}$$

The diatomic potential for AB is given by

$$V_{AB}^{(2)} = \frac{c_0 \exp(-\alpha_{AB} R_1)}{R_1} + \sum_{i=1}^{L} c_i \rho_1^i \ . \tag{135}$$

For BC and CA the expressions are similar. The Rydberg type variables ρ_i are given by

$$\rho_i = R_i \exp(-\beta_i R_i). \tag{136}$$

The three-body term $V_{ABC}^{(3)}$ is written as

$$V_{ABC}^{(3)}(R_1, R_2, R_3) = \sum_{ijk}^{M} d_{ijk} \rho_1^i \rho_2^j \rho_3^k. \tag{137}$$

The long-range potential V_{LR} for H^-–H_2 is expressed as

$$V_{LR} = \frac{qQ(r)P_2(\cos\gamma)}{R^3} - \frac{q^2}{2R^4}[\alpha_0(r) + \alpha_2(r)P_2(\cos\gamma)], \tag{138}$$

where q is the charge on the H^- ion, and $Q(r)$ is the quadrupole moment

$$Q(r) = (a_0 + a_1 r + a_2 r^2 + a_3 r^3 + a_4 r^4 + a_5 r^5)\exp(-a_6 r). \tag{139}$$

$\alpha_0(r)$ and $\alpha_2(r)$ are defined by

$$\alpha_0(r) = \begin{cases} (b_0 + b_1 r + b_2 r^2 + b_3 r^3 + b_4 r^4)\exp(-b_5 r), & 0.4 \leq r \leq 4.0 \text{Bohr} \\ b_6 + b_7 \exp[-2.6(r - 4.0)], & r > 4.0 \text{ Bohr} \end{cases} \tag{140}$$

$$\alpha_2(r) = c_0 \exp[-c_1(r - 3.2)^2]. \tag{141}$$

All parameters are fitted to *ab initio* energies with a least-squared-method.

The original fortran code of Panda and Sathyamurthy has supplied the derivatives of the potential energy with respect to the three internuclear distances (r_1, r_2, r_3). But these derivatives were not correct because they did not include the long-range potential part. We modified this in order to get the correct derivatives.

Ayouz et al. [54] determined a new potential energy surface for the electronic ground state of the H_3^- ion. The coupled-electron pair approximation (CEPA-2) method [138] was used in the calculation of this PES. A large basis set, AV5Z with *spdf/g* basis functions, and in comparison to the work of Stärck and Meyer, and Panda and Sathyamurthy a much larger number of geometries (3024) were used in these studies.

In the present work these three PESs (SM, PS, and AY) have been investigated in detail. The three potential energy surfaces are plotted in Fig. 7 using Jacobi coordinates R and r for three angles $\gamma = 0, 45, 90°$. Values for PS-PES and AY-PES are plotted at the three top panels in Fig. 8. The left top panel in Fig. 8 shows for $\gamma = 0°$ and r values in the range of 1.2–2.4 Bohr that the difference of the PESs is lower than 0.8 eV. For $r > 2.4$ Bohr deviation is larger than 0.8 eV. For r in the range of 1.4–2.4 Bohr and R in the range of 4.5–6.5 Bohr, $\triangle V$(PS-AY) is lower than 0.2 eV. The middle top panel shows that for r between 1.2 and 2.4 Bohr and R in the range of 2–10 Bohr, $\triangle V$(PS-AY) is between 0.2 and 0.8 eV. For $\gamma = 90°$ deviation is similar to $\gamma = 45°$. $\triangle V$(PS-AY) is mostly larger than 0.2 eV; some deviations are even larger than 1.6 eV. For r larger than 2.5 Bohr $\triangle V$(PS-AY) is larger than 3.2 eV. All these results indicate that there are large differences between PS-PES and AY-PES.

A comparison of PS-PES and SM-PES is plotted in the three middle panels in Fig. 8. All these three panels show that the difference of potential energy values are mostly larger than 0.2 eV. $\triangle V$(PS-SM) is smaller than 0.2 eV for $r = [1.3\text{--}2.0]$ Bohr and $R = [3.0\text{--}5.0]$ Bohr. The three bottom panels in Fig. 8 show $\triangle V$(SM-AY) for the angles $\gamma = 0, 45, 90°$. It can be seen that in the interesting part $(1.0 < r < 2.3 \text{ Bohr})$ $\triangle V$(SM-AY) is very small. $\triangle V$(SM-AY) values are lower than 0.1 eV, especially in the area of $r = [1.2\text{--}1.6]$ Bohr, i.e., near the energy minimum of the H_2 curve $\triangle V$(SM-AY) < 0.05 eV. This feature tells us that in these areas the potential energy surface SM-PES and AY-PES are very similar. We see that for $r > 2.5$ Bohr the energy gap is larger than 3.2 eV. This results from an incorrect dissociation energy of AY-PES (see Fig. 9).

From all the discussion above we can get the conclusion that the most interesting part of the potential for SM-PES and AY-PES is very similar.

5 Results

5.1 $H^+ + H_2(v, j)$

5.1.1 Non-Born-Oppenheimer investigations

In order to understand the influence of non-Born-Oppenheimer effects in reactive processes of triatomic systems, the collision of $H^+ + H_2$ was chosen as a typical example, where one has avoided crossings and conical intersections. The change from the electronic ground state to the first excited state leads to a charge transfer.

The non-Born-Oppenheimer semi-classical trajectory method of Truhlar and coworkers (adiabatic and non-adiabatic trajectory program ANT07 [139]) was used to investigate the collision of a proton with different vibrational excited H_2-molecules ($H_2(v$=0–5, j=0)). For a better understanding of the non-adiabatic effects, first adiabatic single-surface ground state calculations were performed using the CTAMYM program [147]. These results were then compared with the multi-surface calculations.

The trajectory calculations were performed by running a batch of 10^4 trajectories at collision energies 0.01–3.0 eV. The integration step-size in the trajectories was chosen to be 0.05 fs. This guarantees conservation of total energy and total angular momentum. The maximum impact parameter b_{max} was chosen as 4.1 Å. Each trajectory was integrated until one atom was separated from the other two by at least 20 Å. The final electronic state and the molecular arrangement could therefore be assigned unambiguously, and product branching probabilities were obtained by counting trajectories. The 3×3 DIM-potential of Kamisaka et al. [35] (KBNN) was used for both adiabatic and non-adiabatic calculations.

5.1.2 Reaction probabilities

The reaction probabilities for the collisions $H^+ + H_2(v$=0–5, j=0) $\rightarrow H_2(v',j') + H^+$ were investigated in three different ways:

Method AD1: Ground state adiabatic surface calculations using the program CTAMYM [147]. The reaction probabilities are listed in Tab. 1.

Method AD2: Ground state adiabatic surface calculations using the program ANT07. The reaction probabilities are listed in Tab. 2.

Method NAD: Non-adiabatic multi-surface calculations using the program ANT07. The reaction probabilities are listed in Tabs. 3–8.

In the case of NAD-investigations several different product channels have to be taken into account: reactive charge transfer (R-CT), nonreactive charge transfer (NR-CT), reactive no-charge-transfer (R-NCT), and nonreactive no-charge-transfer (NR-NCT). In the following, three product channels are considered:

R-CT: $H^+ + H_2(v, j) \rightarrow H_2^+(v', j') + H$ ($A^+ + BC \rightarrow AB^+ + C$)

NR-CT: $H^+ + H_2(v, j) \rightarrow H + H_2^+(v', j')$ ($A^+ + BC \rightarrow A + BC^+$)

R-NCT: $H^+ + H_2(v, j) \rightarrow H_2(v', j') + H^+$ ($A^+ + BC \rightarrow AB + C^+$)

Table 1: AD1: Total reaction probabilities P^{AD1} for different initial vibrational states $H^+ +$ $H_2(v{=}0{-}5, j{=}0) \rightarrow H_2(v',j') + H^+$ on the lowest adiabatic KBNN-PES using the CTAMYM code.[1]

$E_{coll}[\text{eV}]$	$P(v=0)$	$P(v=1)$	$P(v=2)$	$P(v=3)$	$P(v=4)$	$P(v=5)$
0.01	0.6402	0.6573	0.6437	0.6184	0.6391	0.5754
0.1	0.6539	0.6111	0.5675	0.5334	0.5128	0.4915
0.2	0.5416	0.5334	0.5090	0.4911	0.4927	0.4734
0.3	0.4234	0.4000	0.3983	0.4384	0.4556	0.4446
0.4	0.3249	0.3112	0.3368	0.3786	0.4142	0.4038
0.5	0.2543	0.2540	0.2918	0.3262	0.3613	0.3657
0.6	0.2086	0.2129	0.2518	0.2842	0.3115	0.3202
0.7	0.1687	0.1826	0.2145	0.2502	0.2773	0.2845
0.8	0.1410	0.1573	0.1839	0.2227	0.2506	0.2498
0.9	0.1263	0.1357	0.1585	0.1933	0.2214	0.2236
1.0	0.1094	0.1172	0.1348	0.1697	0.1992	0.1995
1.1	0.0933	0.1012	0.1231	0.1480	0.1743	0.1771

[1] $N_{tot} = 10000$, $b = 4.1$ Å, $\Delta t = 0.05$ fs

In Tabs. 3–8, P_{CT}^R is the reaction probability for the channel R-CT; P_{CT}^{NR} is the reaction probability for the channel NR-CT; P_{NCT}^R is the reaction probability for the channel R-NCT; $P_{CT,NCT}^R$ is the sum over the reaction probabilities P_{CT}^R and P_{NCT}^R; $P_{CT}^{R,NR}$ is the sum over the reaction probabilities P_{CT}^R and P_{CT}^{NR}; P_{tot}^R is the total reaction probability, i.e., $P_{NCT}^R + P_{CT}^{R,NR}$.

In order to get a better overview of the differences in the reaction probabilities, obtained from the different methods (AD1, AD2, and NAD) the reaction probabilities had been plotted in Figs. 10–12.

Fig. 10 shows the reaction probabilities P^{AD1}, P^{AD2}, $P_{CT,NCT}^R$ for the methods AD1, AD2, and NAD. From Fig. 10 it can be seen that when the systems is at low energy, the reaction probabilities P^{AD1}, P^{AD2}, $P_{CT,NCT}^R$ are the same. For increasing collision energy the reaction probabilities become different. For $v > 3$ the difference becomes even larger at low collision energies (see panels E($v = 4$) and F($v = 5$) in Fig. 10). The reaction probabilities calculated with the adiabatic single-surface using the two programs CTAMYM and ANT07 deviate from each other at the low collision energy $E_{coll} = 0.01$ eV (more than 40%, see panel E($v = 4$) in Fig. 10).

Fig. 11 shows the reaction probabilities P^{AD1}, P^{AD2}, and P_{tot}^R for the three methods AD1, AD2, and NAD. P_{tot}^R is the total reaction probability calculated with method NAD, i.e., the sum over P_{CT}^R, P_{CT}^{NR}, and P_{NCT}^R. From Fig. 11 one can see that in the case that the reactant H_2 starts with a low vibrational state and at low collision energy, the reaction is adiabatic. When the reactant H_2 starts with a high vibrational state (normally $v > 3$), the total reaction

Table 2: AD2: Total reaction probabilities P^{AD2} for different initial vibrational states $H^+ + H_2(v{=}0{-}5,\ j{=}0) \rightarrow H_2(v', j') + H^+$ on the lowest adiabatic KBNN-PES using the ANT07 code.[1]

E_{coll}[eV]	$P(v=0)$	$P(v=1)$	$P(v=2)$	$P(v=3)$	$P(v=4)$	$P(v=5)$
0.01	0.5270	0.3547	0.2926	0.2502	0.2135	0.5148
0.1	0.6539	0.6111	0.5675	0.5334	0.5128	0.4915
0.2	0.5416	0.5334	0.5090	0.4911	0.4927	0.4734
0.3	0.4234	0.4000	0.3983	0.4384	0.4556	0.4446
0.4	0.3249	0.3112	0.3368	0.3786	0.4142	0.3986
0.5	0.2543	0.2540	0.2918	0.3262	0.3613	0.3657
0.6	0.2086	0.2129	0.2518	0.2842	0.3115	0.3202
0.7	0.1687	0.1826	0.2145	0.2502	0.2773	0.2845
0.8	0.1410	0.1573	0.1839	0.2227	0.2506	0.2498
0.9	0.1263	0.1357	0.1585	0.1933	0.2214	0.2236
1.0	0.1094	0.1172	0.1348	0.1697	0.1992	0.1995
1.1	0.0933	0.1012	0.1231	0.1480	0.1743	0.1771

[1] $N_{tot}= 10000$, $b = 4.1$ Å, $\Delta t = 0.05$ fs

Table 3: NAD: Reaction probabilities P for the $H^+ + H_2(v{=}0, j{=}0)$ collision using the diabatic representation of the KBNN-PES.[1]

E_{coll}[eV]	$P^R_{CT,NCT}$	P^R_{NCT}	$P^{R,NR}_{CT}$	P^R_{tot}	P^{NR}_{CT}
0.01	0.5252	0.5252	0.0000	0.5252	0.0000
0.1	0.6487	0.6487	0.0000	0.6487	0.0000
0.2	0.5402	0.5402	0.0000	0.5402	0.0000
0.3	0.4232	0.4232	0.0000	0.4232	0.0000
0.4	0.3249	0.3249	0.0000	0.3249	0.0000
0.5	0.2543	0.2543	0.0000	0.2543	0.0000
0.6	0.2086	0.2086	0.0000	0.2086	0.0000
0.7	0.1687	0.1687	0.0000	0.1687	0.0000
0.8	0.1410	0.1410	0.0000	0.1410	0.0000
0.9	0.1263	0.1263	0.0000	0.1263	0.0000
1.0	0.1094	0.1094	0.0000	0.1094	0.0000
1.1	0.0933	0.0933	0.0000	0.0933	0.0000
1.5	0.0658	0.0658	0.0000	0.0658	0.0000
2.0	0.0460	0.0453	0.0011	0.0464	0.0004
2.5	0.0335	0.0293	0.0074	0.0367	0.0032
3.0	0.0268	0.0212	0.0109	0.0321	0.0053

[1] $N_{tot} = 10000$, $b = 4.1$ Å, $\Delta t = 0.05$ fs, ANT07

Table 4: NAD: Reaction probabilities P for the $H^+ + H_2(v{=}1, j{=}0)$ collision using the diabatic representation fo the KBNN-PES.[1]

E_{coll}[eV]	$P^R_{CT,NCT}$	P^R_{NCT}	$P^{R,NR}_{CT}$	P^R_{tot}	P^{NR}_{CT}
0.01	0.3547	0.3547	0.0000	0.3547	0.0000
0.1	0.6111	0.6111	0.0000	0.6111	0.0000
0.2	0.5333	0.5333	0.0000	0.5333	0.0000
0.3	0.4000	0.4000	0.0000	0.4000	0.0000
0.4	0.3112	0.3112	0.0000	0.3112	0.0000
0.5	0.2539	0.2539	0.0000	0.2539	0.0000
0.6	0.2129	0.2129	0.0000	0.2129	0.0000
0.7	0.1826	0.1826	0.0000	0.1826	0.0000
0.8	0.1573	0.1573	0.0000	0.1573	0.0000
0.9	0.1357	0.1357	0.0000	0.1357	0.0000
1.0	0.1172	0.1172	0.0000	0.1172	0.0000
1.1	0.1012	0.1012	0.0000	0.1012	0.0000
1.5	0.0694	0.0676	0.0030	0.0706	0.0011
2.0	0.0475	0.0418	0.0141	0.0559	0.0084
2.5	0.0348	0.0283	0.0152	0.0435	0.0087

[1] $N_{tot} = 10000$, $b = 4.1$ Å, $\Delta t = 0.05$ fs, ANT07

Table 5: NAD: Reaction probabilities P for the $H^+ + H_2(v{=}2, j{=}0)$ collision using the diabatic representation of KBNN-PES.[1]

E_{coll}[eV]	$P^R_{CT,NCT}$	P^R_{NCT}	$P^{R,NR}_{CT}$	P^R_{tot}	P^{NR}_{CT}
0.01	0.2926	0.2926	0.0000	0.5675	0.0000
0.1	0.5675	0.5675	0.0000	0.5675	0.0000
0.2	0.5090	0.5090	0.0000	0.5090	0.0000
0.3	0.3983	0.3983	0.0000	0.3983	0.0000
0.4	0.3368	0.3368	0.0000	0.3368	0.0000
0.5	0.2918	0.2918	0.0000	0.2918	0.0000
0.6	0.2518	0.2518	0.0000	0.2518	0.0000
0.7	0.2145	0.2145	0.0000	0.2145	0.0000
0.8	0.1841	0.1841	0.0000	0.1841	0.0000
0.9	0.1583	0.1579	0.0015	0.1594	0.0011
1.0	0.1349	0.1334	0.0031	0.1365	0.0016
1.1	0.1230	0.1200	0.0087	0.1287	0.0050

[1] $N_{tot} = 10000$, $b = 4.1$ Å, $\Delta t = 0.05$ fs, ANT07

Table 6: NAD: Reaction probabilities P for the $H^+ + H_2(v=3, j=0)$ collision using the diabatic representation of the KBNN-PES.[1]

$E_{coll}[\text{eV}]$	$P^R_{CT,NCT}$	P^R_{NCT}	$P^{R,NR}_{CT}$	P^R_{tot}	P^{NR}_{CT}
0.01	0.2502	0.2502	0.0000	0.2502	0.0000
0.1	0.5334	0.5334	0.0000	0.5334	0.0000
0.2	0.4911	0.4911	0.0000	0.4911	0.0000
0.3	0.4384	0.4381	0.0004	0.4385	0.0001
0.4	0.3784	0.3776	0.0022	0.3798	0.0014
0.5	0.3256	0.3230	0.0074	0.3304	0.0048
0.6	0.2833	0.2782	0.0152	0.2934	0.0101
0.7	0.2489	0.2402	0.0257	0.2659	0.0170
0.8	0.2200	0.2079	0.0378	0.2457	0.0257
0.9	0.1918	0.1794	0.0458	0.2252	0.0334
1.0	0.1676	0.1542	0.0558	0.2100	0.0426
1.1	0.1449	0.1302	0.0632	0.1934	0.0485

[1] $N_{tot} = 10000$, $b = 4.1$ Å, $\Delta t = 0.05$ fs, ANT07

Table 7: NAD: Reaction probabilities P for $H^+ + H_2(v=4, j=0)$ collision using the diabatic representation of the KBNN-PES.[1]

$E_{coll}[\text{eV}]$	$P^R_{CT,NCT}$	P^R_{NCT}	$P^{R,NR}_{CT}$	P^R_{tot}	P^{NR}_{CT}
0.01	0.1876	0.182	0.0382	0.2202	0.0326
0.1	0.4451	0.4311	0.1030	0.5341	0.0890
0.2	0.4420	0.4208	0.1212	0.5420	0.1000
0.3	0.4012	0.3733	0.1455	0.5188	0.1171
0.4	0.3700	0.3413	0.1515	0.4928	0.1228
0.5	0.3282	0.2979	0.1893	0.4872	0.1598
0.6	0.2973	0.2658	0.2120	0.4778	0.1805
0.7	0.2594	0.2272	0.2436	0.4708	0.2114
0.8	0.2296	0.1978	0.2688	0.4666	0.2377
0.9	0.2043	0.1743	0.2894	0.4637	0.2594
1.0	0.1821	0.1535	0.3020	0.4555	0.2734
1.1	0.1664	0.1327	0.3221	0.4548	0.2884

[1] $N_{tot} = 10000$, $b = 4.1$ Å, $\Delta t = 0.05$ fs, ANT07

Table 8: The reaction probabilities P for $H^+ + H_2(v=5, j=0)$ collision using the diabatic representation of the KBNN-PES.[1]

E_{coll}[eV]	$P^R_{CT,NCT}$	P^R_{NCT}	$P^{R,NR}_{CT}$	P^R_{tot}	P^{NR}_{CT}
0.01	0.3159	0.2785	0.2365	0.5150	0.1991
0.1	0.3047	0.2654	0.2703	0.5357	0.2310
0.2	0.2928	0.2481	0.2917	0.5398	0.2470
0.3	0.2746	0.2305	0.3058	0.5363	0.2617
0.4	0.2448	0.2015	0.3186	0.5201	0.2753
0.5	0.2189	0.1758	0.3380	0.5138	0.2949
0.6	0.1970	0.1559	0.3452	0.5011	0.3041
0.7	0.1844	0.1411	0.353	0.4941	0.3097
0.8	0.1671	0.1272	0.3756	0.5028	0.3357
0.9	0.1541	0.1140	0.3775	0.4915	0.3374
1.0	0.1307	0.0978	0.3926	0.4904	0.3597
1.1	0.121'	0.082	0.402	0.484	0.363

[1] $N_{tot} = 10000$, $b = 4.1$ Å, $\Delta t = 0.05$ fs, ANT07

probability calculated with a multi-surface is higher than that calculated with a single surface. This means that the adiabatic calculation is not good enough to provide correct reaction probabilities for high energies. In this case, non-adiabatic effects should be considered.

The reaction probabilities $P^{R,NR}_{CT}$ are listed in Tab. 9 and plotted in Fig. 12. From Tab. 9 and Fig. 12 one can see that the products can be found on the first excited surface only when the system is at higher energies. In the case that the system is at low collision energies, the products prefer to stay on the ground adiabatic surface. In the case that the H_2 reactant starts with rovibrational state $v=0$, $j=0$, and the H_2^+ product is searched for, the collision energy should not be lower than 2.0 eV. If the reactant H_2 starts with a high initial rovibrational state, e.g. $v=5$, $j=0$, the H_2^+ product can be found even at low collision energies.

5.1.3   The surface hopping analysis

Figs. 13, 14, 15 show the number (index) of the three potential energy surfaces and the three internuclear distances (r(A-B), r(A-C), r(B-C)). The number 1 (on the right side of each panel) is the index of the ground adiabatic potential surface, the numbers 2 and 3 are the indices of the first and second excited adiabatic potential surfaces. Panel A in Fig. 13 shows the reactant $H_2(v=3)$ and the collision energy $E_{coll} = 0.90$ eV; adiabatic surface hopping occurs once at nearly the collision time $T = 350$ fs. The final product is on the first excited surface, namely, this reaction is NR-CT ($H^+ + H_2 \rightarrow H + H_2^+$). Panels B and E in Fig. 13 show the reactant $H_2(v=4)$ and E_{coll} is 0.70 and 1.00 eV; from these two panels one can see that the product is $H_2(v' < 4)$. There is no surface hopping in the whole process. These two reactions are

Table 9: The reaction probabilities $P_{CT}^{R,NR}(v)$ for the charge transfer process $H^+ + H_2(v=0\text{--}5,$ $j=0) \to H + H_2^+\,(^2\Sigma_g^+)(v',j')$.

E_{coll}[eV]	$P_{CT}^{R,NR}(v=0)$	$P_{CT}^{R,NR}(v=1)$	$P_{CT}^{R,NR}(v=2)$	$P_{CT}^{R,NR}(v=3)$	$P_{CT}^{R,NR}(v=4)$	$P_{CT}^{R,NR}(v=5)$
0.01	0.0000	0.0000	0.0000	0.0000	0.0382	0.2365
0.1	0.0000	0.0000	0.0000	0.0000	0.1030	0.2703
0.2	0.0000	0.0000	0.0000	0.0000	0.1212	0.2917
0.3	0.0000	0.0000	0.0000	0.0004	0.1455	0.3058
0.4	0.0000	0.0000	0.0000	0.0022	0.1515	0.3186
0.5	0.0000	0.0000	0.0000	0.0074	0.1893	0.3380
0.6	0.0000	0.0000	0.0000	0.0152	0.2120	0.3452
0.7	0.0000	0.0000	0.0000	0.0257	0.2436	0.3530
0.8	0.0000	0.0000	0.0000	0.0378	0.2688	0.3756
0.9	0.0000	0.0000	0.0015	0.0458	0.2894	0.3775
1.0	0.0000	0.0000	0.0031	0.0558	0.3020	0.3926
1.1	0.0000	0.0000	0.0087	0.0632	0.3221	0.4020
1.5	0.0000	0.0030	———	———	———	———
2.0	0.0011	0.0141	———	———	———	———
2.5	0.0074	0.0152	———	———	———	———
3.0	0.0109	———	———	———	———	———

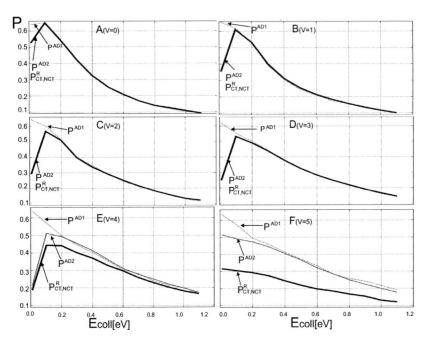

Figure 10: Reaction probabilities P^{AD1}, P^{AD2}, $P^R_{CT,NCT}$ obtained from the methods AD1, AD2, and NAD as a function of collision energy E_{tot} for different initial vibrational states (the initial rotational state is $j = 0$).

NR-NCT ($H^+ + H_2 \rightarrow H^+ + H_2$). Panels C, D, and F in Fig. 13 are similar to each other. The common features in these panels are: first, the collision time of these reactions is short; second, the final vibrational state of the products are greater than 4 resulting in many surface hopping events in these reactions after collision. From panel D in Fig. 13 one can see that the final product is on the first excited potential surface. This reaction belongs to NR-CT. The reactions, which are shown in panels C and F in Fig. 13, belong to NR-NCT.

Fig. 14 shows the reactant $H_2(v=5)$ for $E_{coll} = 0.01$–0.50 eV. Because the initial internuclear distance r(A-B) of $H_2(v=5)$ can be greater than 2.5 Bohr, surface hopping occurs at a low collision energy. In the case of low initial collision energies $E_{coll} < 0.4$ eV, the separated atom can take away some energy from the diatomic molecule, so the final vibrational state of the diatomic molecule will be decreased (see panels G, F, I, J in Fig. 14). There is no surface hopping after the collision. With an increase in E_{coll} the final vibrational state of the diatomic molecule will not be decreased. In this case, there is surface hopping after the collision (see panels K and L in Fig. 14). Panels G and I in Fig. 14 show that the arrangements of the atoms are different between the reactant and the product. These two reactions belong to

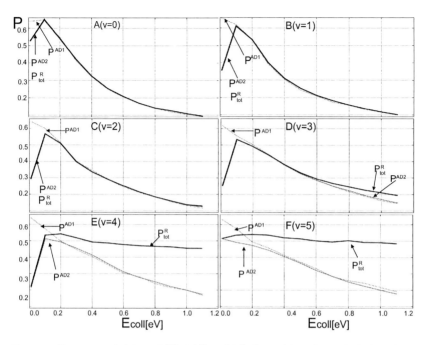

Figure 11: Reaction probabilities P^{AD1}, P^{AD2}, and P^R_{tot} obtained from the methods AD1, AD2, and NAD as a function of collision energy E_{tot} for different initial vibrational states (the initial rotational state is $j = 0$).

R-NCT ($H^+ + H_2 \rightarrow H_2 + H^+$). The reactions, which are shown by panels H, K, J, and L, belong to NR-NCT.

Fig. 15 shows the reactant $H_2(v{=}5)$ for $E_{coll} = 0.60$–1.10 eV. From Fig. 15 one can see that with an increase in E_{coll}, the time that the system stays on the first excited potential surface becomes longer and longer. In these reactions, only one reaction (see panel R in Fig. 15) belongs to R-NCT; the other reactions (see panels M, N, O, P, and Q in Fig. 15) are NR-NCT.

From all the discussions in this section we can conclude that in the case that the reactant H_2 starts with a low initial rovibrational state, and we want to achieve the product on the first excited potential surface, the reactants should be given higher collision energies. In the case that the reactant H_2 starts with the initial rovibrational state $v = 0$, $j = 0$, then the product can be found on the first excited surface only if the collision energies are greater than 2.0 eV. In the case of reactants $H_2(v{=}1$, $j{=}0)$ or $H_2(v{=}2$, $j{=}0)$, the corresponding collision energies should be greater than 1.4 or 0.9 eV in order to get products on the first excited surface (see

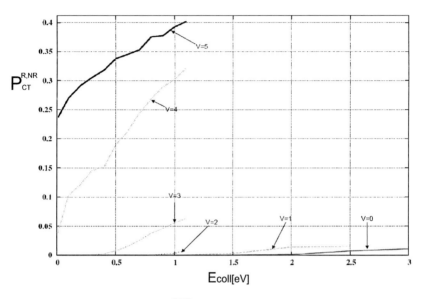

Figure 12: The reaction probability $P_{CT}^{R,NR}$ using method NAD as a function of collision energy E_{coll}. $P_{CT}^{R,NR}$ is sum over the reaction probability for $H^+ + H_2(v=0-5, j=0) \rightarrow H + H_2^+(v', j')$ (NR-CT) and $H^+ + H_2(v=0-5, j=0) \rightarrow H_2^+(v', j') + H$ (R-CT).

Tab. 9).

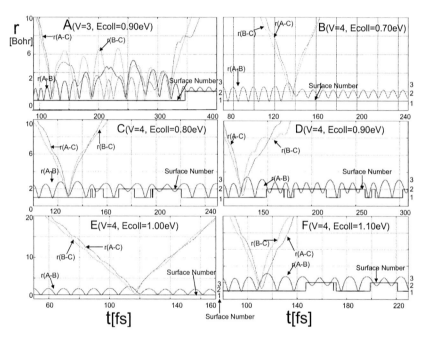

Figure 13: Representative trajectories and adiabatic surface hopping for $H^+ + H_2(v{=}3{-}4, j{=}0)$ in different collision energies. The potential surface number and the internuclear distances r(A-B), r(B-C), and r(A-C) are shown as a function of time.

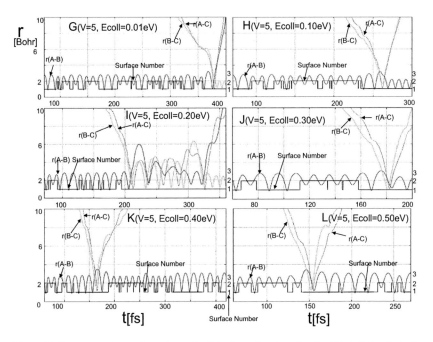

Figure 14: Representative trajectories and adiabatic surface hopping for H^+ + $H_2(v=5, j=0)$ in different lower collision energies. The potential surface number and internuclear distances r(A-B), r(B-C), and r(A-C) are shown as a function of time.

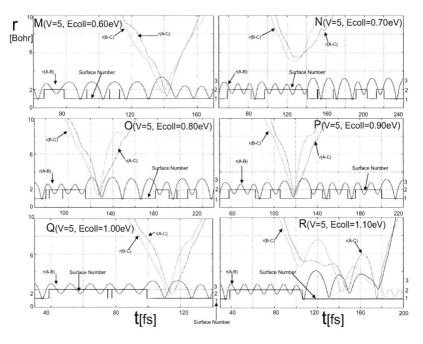

Figure 15: Representative trajectories and adiabatic surface hopping for $H^+ + H_2(v=5, j=0)$ in different higher collision energies. The potential surface number and internuclear distances r(A-B), r(B-C), and r(A-C) are shown as a function of time.

5.2 $H^- + H_2(v,j)$

The study of H^- and its interaction with other species like H_2 is of fundamental interest and also of importance from the point of view of plasmas and interstellar media [73]. Reactions of a hydrogen atom and its negative and positive counterparts with a hydrogen molecule are of fundamental importance as they constitute prototype systems to test *ab initio* theories against experiment. The neutral counterpart H_3 and its isotopic analogs have been particularly studied over the years [144, 145, 146]. These studies have shown that the potential energy surface has a barrier of 0.4167 eV for the exchange reaction [146].

Qualitatively, the potential energy surface (PES) of H_3^- is very similar to that of H_3 in many ways besides the barrier height. However, the PES of H_3^- has special features. First, when the H^- atom is far away from the H_2 molecule, the charge-induced dipole interaction between the two collision parties can lead to a minimum; the depth is only 0.05 eV. Second, there is an additional electron in the H_3^- system. So, such a system has an additional reaction channel, i.e., the electron detachment channel. The electron detachment reaction will occur at around 1.45 eV [50]. This reaction pathway looks formally like:

$$H^- + H_2 \rightarrow [\, H_3^- \,]^{\neq} \rightarrow H + H_2 + e^- \quad .$$

A crossed beam study of the rearrangement reaction $H^- + D_2(v{=}0) \rightarrow HD(v') + D^-$, with collision energies between 0.3 eV and 3 eV, was reported by Zimmer and Linder (1995) [65]. From this experiment we know that the reaction has a threshold at $E_{rel} = 0.42 \pm 0.12$ eV. The cross section rises to a maximum of $2 * 10^{-16}$ cm^2 at 1.5 eV and then rapidly decreases.

5.2.1   Quasi-classical trajectory investigations

The $H^- + H_2(v{=}0{-}5,\ j{=}0) \rightarrow H_2(v',j') + H^-$ reaction was studied using the CTAMYM [147] program for the potential surfaces of Stärck and Meyer (SM-PES) [50] and Panda and Sathyamurthy (PS-PES) [53]. The trajectories start with the correct quantum mechanical rovibrational states. On the other hand, in the analysis part of the trajectory calculations the final vibrational energies can be below zero point energy. This can lead to a small effective reaction barrier.

Initial parameter determinations:
(1) The maximum impact parameter b_{max} was gained by investigating batches of 10^4 trajectories for different fixed impact parameters b. The value of b was stepwise increased. The calculations are performed using the programs CTAMYM [147] and Venus-96 [148] for SM-PES and PS-PES. The initial rovibrational state for the diatomic molecule H_2 is $v = 0$, $j = 0$. The collision energies are $E_{coll} = 0.5$ eV and 1.0 eV. Detailed results are shown in Tab. 10 and 11 and plotted in Fig. 16.   As can be seen from Tab. 10 and 11 and Fig. 16, the maximum impact parameter b_{max} is 1.3 Å for $E_{coll} = 0.5$ eV and SM-PES (CTAMYM and VENUS-96). For $E_{coll} = 1.0$ eV the maximum impact parameter b_{max} is 1.5 Å (CTAMYM) and 1.6 Å (VENUS-96) for SM-PES. For PS-PES and $E_{coll} = 0.5$ eV the maximum impact

Table 10: H^- + H_2: The reaction probabilities P_C, P_V for the SM-PES with $E_{coll} = 0.5$ and 1.0 eV and for the initial state $v = 0$, $j = 0$ (using the program packages CTAMYM and VENUS-96 with 10^4 trajectories).

E_{coll}	0.5 [eV]		1.0 [eV]	
$b[\text{Å}]$	P_C	P_V	P_C	P_V
0.1	0.7169	0.7158	0.8408	0.8381
0.2	0.7073	0.7045	0.8451	0.8418
0.3	0.6841	0.6841	0.8496	0.8454
0.4	0.6421	0.6407	0.8455	0.8454
0.5	0.6055	0.6019	0.8344	0.8455
0.6	0.5549	0.5574	0.8174	0.8355
0.7	0.4965	0.4990	0.7847	0.8148
0.8	0.4286	0.4295	0.7438	0.7859
0.9	0.3498	0.3504	0.6947	0.7469
1.0	0.2718	0.2717	0.6274	0.6999
1.1	0.1873	0.1865	0.5366	0.6359
1.2	0.1032	0.1037	0.3920	0.5406
1.3	0.0233	0.0232	0.2605	0.4000
1.4	0.0000	0.0000	0.1412	0.2709
1.5	0.0000	0.0000	0.0462	0.1423
1.6	0.0000	0.0000	0.0000	0.0481
1.7	0.0000	0.0000	0.0000	0.0000

P_C : using CTAMYM; P_V : using VENUS-96

parameter b_{max} is 1.2 Å; in case of $E_{coll} = 1.0$ eV $b_{max} = 1.4$ Å has to be used. From Tab. 10 and 11, and Fig. 16 we can see that the maximum impact parameter b_{max} increases with an increase in the collision energy. We determined the reaction probabilities P and the reaction cross sections σ for different maximum impact parameters b_{max} (see Tab. 12) and obtained similar reaction cross sections. In this way, the maximum impact parameter b_{max} was specified for the other initial vibrational states $v = 1$–5 (see Tab. 13). (2) The initial internal energies, turning points, and vibrational half-periods of H_2 (using SM-PES and PS-PES) are shown in Tab. 14 and 15.

Reaction probabilities:
In the present work the reaction probabilities for initial vibrational states $v = 0$–5 using PS-PES and SM-PES were calculated. The quasi-classical trajectory (QCT) calculations were performed by running batches of 10^5 trajectories at 0.01–3.5 eV collision energy. The integration step-size in the trajectories was chosen to be 0.05 fs. This guarantees conservation of total energy and total angular momentum. The trajectories were started at a distance between the

Table 11: H^- + H_2: The reaction probabilities P_C, P_V for the PS-PES with E_{coll}=0.5 and 1.0 eV and with the initial state $v = 0$, $j = 0$ (using the program packages CTAMYM and VENUS-96).

E_{coll}	0.5 [eV]		1.0 [eV]	
b [Å]	P_C	P_V	P_C	P_V
0.1	0.3891	0.5666	0.7829	0.7968
0.2	0.4031	0.5555	0.7760	0.7946
0.3	0.3675	0.5394	0.7645	0.7889
0.4	0.3385	0.5138	0.7452	0.7770
0.5	0.2990	0.4834	0.7209	0.7551
0.6	0.2508	0.4398	0.6829	0.7222
0.7	0.1883	0.3824	0.6336	0.6786
0.8	0.1350	0.3140	0.5754	0.6330
0.9	0.0856	0.2472	0.5077	0.5722
1.0	0.0429	0.1734	0.4222	0.4971
1.1	0.0187	0.1009	0.3186	0.4055
1.2	0.0013	0.0331	0.1768	0.2755
1.3	0.0000	0.0000	0.0800	0.1188
1.4	0.0000	0.0000	0.0109	0.0197
1.5	0.0000	0.0000	0.0000	0.0000

P_C : using CTAMYM; P_V : using VENUS-96

incoming atom and the center-of-mass of the diatomic molecule of 13 Å. These initial parameters are the same for the case of the H^- + $D_2(v$=0–5, j=0–1), D^- + $H_2(v$=0–5, j=0–1), D^- + $D_2(v$=0–5, j=0–1), H^- + HD(v=0–1, j=0–1), and D^- + HD(v=0–1, j=0–1) reactions. Detailed results are shown in Fig. 17. The reaction probabilities using PS-PES are shown in the left panels, and those using SM-PES are shown in the right panels. The important part is enlarged in the same panel. The maximum impact parameter b_{max} [Å] is shown.

For the initial vibrational state $v = 0$ (see the panels $a1$ and $a2$ in Fig. 17, and panel d in Fig. 18) the reaction probability P starts to increase beyond $E_{coll} > 0.3$ eV. Total reaction probability (P_{tot}) reaches its maximum when the collision energy reaches 1.5/1.2 eV for PS-PES/SM-PES and then decreases with an increase in collision energy. P_{tot} for PS-PES and SM-PES shows different maxima. At low collision energies P_{tot} for PS-PES is lower than for SM-PES. The state-to-state results show that the largest reaction probability is found for the H_2 product in the final vibrational state $v' = 0$. The H_2 product with the final vibrational state $v' = 1$ will be produced for collision energies $E_{coll} > 0.5$ eV; the probability reaches its maximum for $E_{coll} = 2.0$ eV. The products with higher final vibrational states can be observed only for higher collision energies.

Panels $b1$ and $b2$ in Fig. 17 show the reaction probabilities P for the collision H^- with $H_2(v$=1,

Table 12: Reaction probabilities P and reaction cross sections σ for H^- + $H_2(v{=}0, j{=}0)$ using the potential energy surface SM and the CTAMYM program for $E_{coll} = 1.0$ eV.

b_{max} [Å]	P	σ [Å^2]
1.6	0.46038	3.70072 (100000 trajectories)
1.7	0.40827	3.70489 (100000 trajectories)
3.0	0.131405	3.71351 (400000 trajectories)
3.0	0.13037	3.68426 (100000 trajectories)
4.1	0.06943	3.66475 (100000 trajectories)
4.1	0.07060	3.72627 (900000 trajectories)

Table 13: The maximum impact parameter b_{max} for H^- + $H_2(v{=}0\text{--}5,j{=}0)$.

v	$v = 0$	$v = 1$	$v = 2$	$v = 3$	$v = 4$	$v = 5$
b_{max} [Å]	2.0	2.2	10.2	10.3	10.6	10.8

Table 14: Internal energies, turning points, and vibrational half-periods of H_2 for different rovibrational (v,j) states (using SM-PES).

v	j	$E_{v,j}$[cm^{-1}]	r_- [Å]	r_+ [Å]	$\tau[fs]$
0	0	2170.466	0.633568	0.884114	3.9078109
0	1	2288.792	0.634635	0.885349	3.9128834
1	0	6324.044	0.570934	1.013968	4.1239307
1	1	6436.612	0.571954	1.01530	4.1304337
2	0	10253.076	0.534689	1.120114	4.3690675
2	1	10359.922	0.535683	1.12154	4.3751542
3	0	13954.735	0.508917	1.219146	4.6473761
3	1	14055.848	0.509894	1.22065	4.6548451
4	0	17426.018	0.489099	1.316244	4.9686883
4	1	17521.355	0.490065	1.31789	4.9776270
5	0	20663.851	0.473215	1.414357	5.3431723
5	1	20753.343	0.474175	1.41615	5.3531415

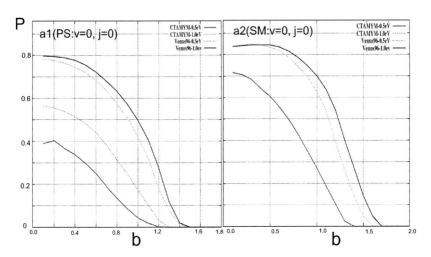

Figure 16: Comparison of different reaction probabilities P as a function of the impact parameter b [Å] for the PESs PS and SM with 10^4 trajectories.

j=0) on PS-PES and SM-PES. P starts to increase at low collision energy ($E_{coll} = 0.01$ eV). For SM-PES and collision energy $E_{coll} \approx 0.4$ eV the P_{tot} reaches a maximum value of ~48%, and then decreases with increasing E_{coll}. For PS-PES, the maximum value of the reaction probability is ~51%, the corresponding collision energy is 0.1 eV. The different vibrational state product distributions show that $H_2(v'=1)$ is the main product at low collision energies. With an increase in E_{coll} the H_2 product with the vibrational state $v' = 0$ becomes more and more favored. For PS-PES the $H_2(v'=1)$ is the most favored product when E_{coll} are in the range of 0.01–0.9 eV. For SM-PES and $E_{coll} = 0.01$–0.3 eV, the main H_2 product has the final vibrational state $v' = 1$. For $E_{coll} > 0.3$ eV the main product $H_2(v'=0)$ is favored. The reaction probabilities are different when the H_2 reactant starts with the initial vibrational state $v = 1$ and $v = 0$: (1) The reaction probabilities for the $H_2(v=1)$ reactant are much higher than for the $H_2(v=0)$ reactant. (2) The final vibrational state distributions for the products are different.

For other initial vibrational states ($v = 2$–5) the features of the reaction probabilities are similar (see panels $c1$, $c2$, $d1$, $d2$, $e1$, $e2$, $f1$ and $f2$ in Fig. 17). All these reactions have the maximum reaction probability reached for collision energies $E_{coll} \approx 0.01$ eV; then the reaction probabilities decrease rapidly with increasing E_{coll}. For $E_{coll} > 0.5$ eV the reaction probability decreases slowly with an increase in E_{coll}. For the H_2 reactant with initial vibrational state $v = 2$, the main vibrational states of the products are $v' = 0, 1, 2$. For $E_{coll} = 0.01$ eV, the highest reaction probabilities for the H_2 products are gained for the final vibrational state $v' = 1$; similar features were calculated for PS-PES and SM-PES. For PS-PES the second and the

Table 15: Internal energies, turning points, and vibrational half-periods of H_2 for different rovibrational (v,j) states (using PS-PES).

v	j	$E_{v,j}[\text{cm}^{-1}]$	r_- [Å]	r_+ [Å]	$\tau[fs]$
0	0	2177.105	0.634755	0.88479	3.8983459
0	1	2295.095	0.635807	0.88602	3.9037283
1	0	6332.546	0.572421	1.01553	4.1302492
1	1	6444.565	0.573432	1.01686	4.1360314
2	0	10250.778	0.536292	1.12242	4.3849765
2	1	10357.021	0.537283	1.12385	4.3910821
3	0	13937.241	0.510476	1.22188	4.6673761
3	1	14037.825	0.511457	1.22341	4.6746746
4	0	17394.291	0.490475	1.31900	4.9858152
4	1	17489.288	0.491452	1.32065	4.9951029
5	0	20623.042	0.474298	1.41654	5.3517626
5	1	20712.465	0.475276	1.41833	5.3615919

third important product is H_2 with final vibrational states $v' = 2$ and $v' = 0$, but for SM-PES, the second important product is H_2 with a final vibrational state $v' = 0$. The product $H_2(v'=2)$ is not important at low collision energies. If the reactant H_2 starts with the initial vibrational state $v = 3$, the highest reaction probability is found for the product $H_2(v'=2)$. The other products $H_2(v'=0,1,3)$ have lower probabilities using PS-PES. For SM-PES the most important product is $H_2(v'=2)$; the products $H_2(v'=0,1)$ have lower probabilities. At low collision energies the product $H_2(v'=3)$ have quite low probabilities. For the reactant $H_2(v=4,5)$ the important products are H_2 with the final vibrational states $v' = 0$, 1, 2, 3, 4. In these products, the largest reaction probability is found for H_2 with the final vibrational state $v' = v-1$ for PS-PES. For SM-PES and the reactant $H_2(v=4, j=0)$, the important products are H_2 with the final vibrational states $v' = 0$, 1, 2, 3. For $H_2(v=5, j=0)$, the important products are H_2 with the final vibrational states $v' = 0$, 1, 2, 3, 4. In these two cases, the largest reaction probability was found for H_2 with the final vibrational state $v' = v-2$.

In Fig. 17, we can see that, at low collision energies, the total reaction probabilities using SM-PES are much higher than those using PS-PES. The vibrational state distributions of the products are different for PS-PES and SM-PES.

Reaction cross sections:

The reaction cross sections σ for H^- + $H_2(v=0–5, j=0)$ was calculated for collision energies E_{coll} in the range of 0.01–3.5 eV. Results for σ $[10^{-16}$ cm$^2]$ are plotted in Fig. 18; panels $b1$ ($b2$) are the enlargement of panels $a1$ ($a2$). In order to compare the different reaction cross sections for PS-PES and SM-PES, the reaction probabilities and reaction cross sections σ for the H^- + $H_2(v=0–1, j=0)$ reaction are plotted in the same panel (c and d).

Fig. 18 demonstrates that the total reaction cross section σ for the reactant $H_2(v=0, j=0)$

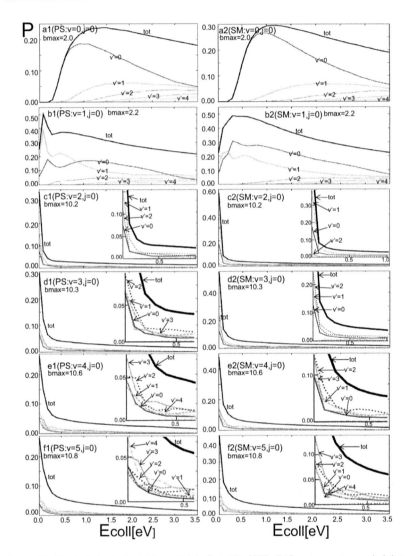

Figure 17: $H^- + H_2(v{=}0{-}5,\ j{=}0) \rightarrow H_2(v',j') + H^-$ (CT): Different reaction probabilities P (P_{tot}, $P_{v'}$: see abbreviation tot or $v' = \cdots$) as a function of collision energy E_{coll} for various initial rovibrational states v,j and different impact parameter b_{max} using the potential energy surfaces SM and PS ($P_{v'} = \sum_{j'} P(v',j')$, $P_{tot} = \sum_{v'} P_{v'}$).

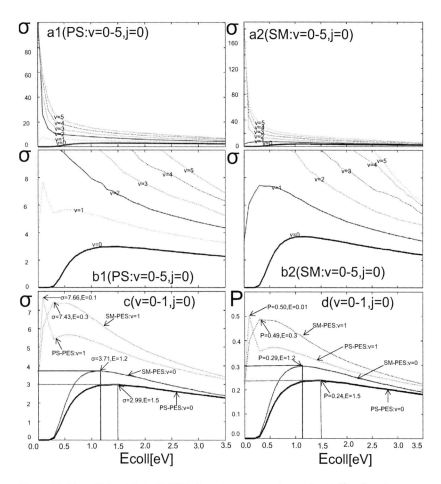

Figure 18: $H^- + H_2(v=0–5, j=0)$ (CT): Reaction cross sections σ [in 10^{-16} cm^2] and reaction probabilities P as a function of collision energy E_{coll} for PS-PES and SM-PES.

increases dramatically with an increase in E_{coll} near the threshold, and the maximum of the reaction cross section is reached for $E_{coll} = 1.5/1.2$ eV for PS-PES/SM-PES. After the maximum, σ decreases slowly with an increase in E_{coll}. This feature is similar for PS-PES and SM-PES, but at low collision energies the cross section σ for SM-PES is much higher than that for PS-PES.

In the case of the $H_2(v=1, j=0)$ reactant the internal energy $E_{v,j}\sim 0.785$ eV is higher than the reaction barrier $E_{bar} = 0.46$ eV. At very low collision energy (0.01 eV), the reaction cross section

Table 16: Parameters used in the wave packet calculations (see the expressions in chapter 3).

N_R, N_r, N_θ	128,128,80	number of grid points for the $J = 0$ reaction
N_R, N_r, N_θ	128,64,32	number of grid points for the $J \neq 0$ reaction
R_{min}, R_{max}	0.001,17.5	extension of the grid in R
r_{min}, r_{max}	0.501,9.0	extension of the grid in r
E_{trans}	1.0	translational energy
R_0	12.0	initial location of the center of the WP
σ_0	0.4	initial width of the WP
N_t	25000	number of time steps
$A, \Delta r_{abs}, \Delta R_{abs}$	0.015,5.0,13.5	parameters for the absorbing potential

σ is already high ($\sigma = 3.82$ and $3.44*10^{-16}$ cm^2 for PS-PES and SM-PES), then σ becomes different with an increase in E_{coll}. For PS-PES σ shows a sharp peak ($\sigma = 7.66*10^{-16}$ cm^2) for $E_{coll} = 0.1$ eV; σ reaches an intermediate minimum at $E_{coll} = 0.3$ eV. After this intermediate minimum σ increases again and reaches another maximum. Beyond this maximum, σ decreases with an increase in E_{coll}. For SM-PES, σ shows a different trend. σ increases with an increase in E_{coll}, and at $E_{coll} = 0.3$ eV, σ reaches its highest value ($\sigma = 7.43*10^{-16}$ cm^2); subsequently, the reaction cross section decreases again with an increase in the collision energy. At low collision energies ($0.2 < E_{coll} < 2.0$ eV), the reaction cross section for PS-PES is much lower than for SM-PES.

For the $H_2(v=2$–$5, j=0)$ reactants the internal energies $E_{v,j}$ are much higher than the reaction barrier; the features of the reaction cross sections are similar to those systems without a reaction barrier, i.e., $\sigma \propto (\frac{1}{E_{kin}})^{\frac{1}{2}}$. The reaction cross sections monotonically decrease with an increase in the collision energy. For PS-PES and $E_{coll} = 0.01$ eV the total reaction cross sections are between 92 and $112*10^{-16}$ cm^2. For SM-PES and $E_{coll} = 0.01$ eV the reaction cross sections lie between 167 and $188*10^{-16}$ cm^2.

5.2.2 Wave packet calculations for $H^- + H_2(v, j) \rightarrow H_2(v', j') + H^-$

This section presents the results of wave packet (WP) (using the real wave packet code of S. Gray [149]) calculations for reaction probabilities. The reaction system is the ion-neutral molecule collision $H^- + H_2(v, j) \rightarrow H_2(v', j') + H^-$ for total angular momenta $J = 0$ and $J \neq 0$. The parameters used in the calculations are listed in Tab. 16.

Time-dependent quantum dynamics of $H^- + H_2(v=0$–$5, j=0) \rightarrow H_2(v', j') + H^-$ *for total angular momentum J = 0:*
Theoretical investigations were performed for the dynamics of reactive scattering processes using time-dependent wave packets for the ion-neutral reaction $H^- + H_2(v=0$–$5, j=0)$ using the potential surfaces PS, SM, and AY. For the inelastic investigations, reactant Jacobi coordinates (PC) are used so that a state-to-state inelastic analysis is possible and energy-dependent total

58

reaction probabilities P_{tot} can be calculated (see right panels in Figs. 19, 20, 21). In order to make sure that the results for the reaction probabilities are converged, firstly, reactant Jacobi coordinates (RC) are used to calculate the total reaction probabilities and then these are compared with those calculated using product Jacobi coordinates (PC) (see left panels in Figs. 19, 20, 21). These panels show that the reaction probabilities, which are calculated using PC-coordinates and RC-coordinates, are the same if the H_2 reactant starts with the initial rovibrational states v=0–2, j=0. But for the $H^- + H_2(v$=3–5, j=0), reaction the reaction probabilities, using different coordinates, deviate. In the following part, we will mainly discuss the title reaction for the H_2 reactant starting with initial rovibrational states v=0–2, j=0.

The right top three panels in Fig. 19 show the total reaction probabilities P_{tot} and state-to-state reaction probabilities $P_{v'}$ for the $H^- + H_2(v$=0, j=0) $\rightarrow$ $H_2(v',j') + H^-$ reaction using PS-PES, SM-PES, and AY-PES. The state-to-state reaction probabilities $P_{v'}$ (summed over rotational states) are calculated using product coordinates (PC) ($J = 0$). For the other reactions the H_2 reactant starts with initial rovibrational states (v=1–5, j=0); similar results are given in the right panels in Figs. 19–21.

In the case of the $H_2(v$=0, j=0) reactant and $J = 0$ the total reaction probabilities are increased immediately at total energy $E_{tot} = 0.5$ eV up to 0.7 eV with a maximum value of ~80%, which is followed by a slower increase up to $E_{tot} = 1.3$ eV with a magnitude of ~87%. For $E_{tot} > 1.5$ eV the total reaction probabilities slowly decrease with an increase in E_{tot}. If the H_2 reactant starts with the initial vibrational state $v = 0$, the highest reaction probabilities are found for the H_2 product in the final vibrational state $v' = 0$. The other vibrational states of the products become populated when the state-to-state probability for $v' = 0$ is decreasing and leveling off (this starts at $E_{tot} \approx 1.7$ eV). In the case of the $H_2(v$=1, j=0) reactant (see right bottom three panels in Fig. 19), the $H_2(v'$=1) product is strongly favored for $E_{tot} = 0.7$–1.6 eV. For the reactants $H_2(v$=2) or $H_2(v$=3), the final vibrational state of the H_2 product is favored for $v' = 2$ at low E_{tot}. For the $H_2(v{\geq}2)$ reactants, accurate reaction probabilities near the threshold are difficult to obtain using the time-dependent wave packet method. At this point, the reaction probabilities lead to a sharp and numerically erroneous peak.

In order to see differences in the total reaction probabilities which are calculated using PS-PES, SM-PES, and AY-PES, results for the total reaction probabilities are plotted in the same panel in Fig. 22. The left panels in Fig. 22 are the total reaction probabilities with total energies up to 3.5 eV. The right panels in Fig. 22 show the total reaction probabilities with E_{tot} up to 2.0 eV. Fig. 22 shows that for the $H_2(v$=0–2, j=0) reactants the reaction probabilities, using SM-PES and AY-PES, are the same for $E_{tot} < 2.0$ eV. For low total energies the reaction probabilities, which one gets from PS-PES, are much lower than those using SM-PES and AY-PES.

Time-dependent quantum dynamics of $H^- + H_2(v$=0, j=0) $\rightarrow H_2(v',j') + H^-$ *for total angular momenta* $J \neq 0$.

Fig. 23 shows the reaction probabilities P for the collision of H^- and $H_2(v$=0, j=0) with total

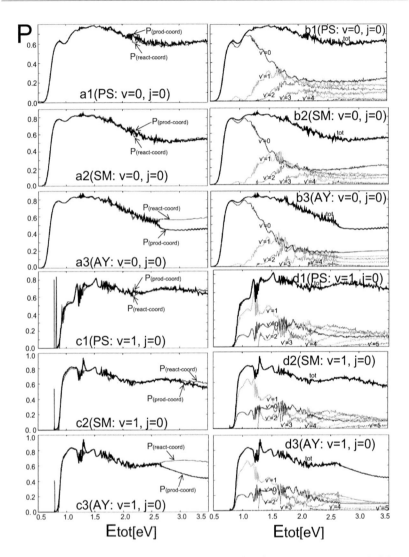

Figure 19: H^- + $H_2(v{=}0\text{--}1,\ j{=}0) \rightarrow H_2(v', j')$ + H^- (WP): Different reaction probabilities (P_{tot}, $P_{v'}$: see abbreviation tot or $v' = \cdot\cdot$) as a function of total energy E_{tot} for various initial vibrational states v and two different Jacobi coordinate systems (RC and PC) using the potential energy surfaces SM, PS, and AY ($P_{v'} = \sum_{j'} P(v', j')$, $P_{tot} = \sum_{v'} P_{v'}$).

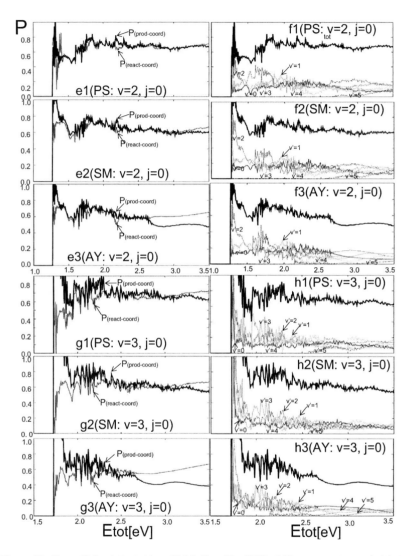

Figure 20: $H^- + H_2(v{=}2{-}3,\ j{=}0) \rightarrow H_2(v', j') + H^-$ (WP) :Different reaction probabilities (P_{tot}, $P_{v'}$: see abbreviation tot or $v' = \cdot\cdot$) as a function of total energy E_{tot} for various initial vibrational states v and two different Jacobi coordinate systems (RC and PC) using the potential energy surfaces SM, PS, and AY ($P_{v'} = \sum_{j'} P(v', j')$, $P_{tot} = \sum_{v'} P_{v'}$).

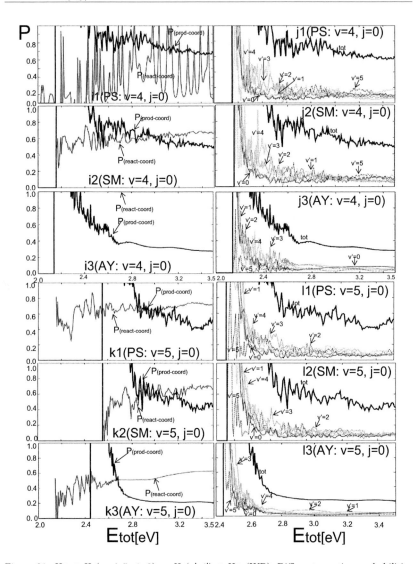

Figure 21: $H^- + H_2(v{=}4\text{--}5,\ j{=}0) \rightarrow H_2(v',j') + H^-$ (WP): Different reaction probabilities (P_{tot}, $P_{v'}$: see abbreviation tot or $v' = \cdots$) as a function of total energy E_{tot} for various initial vibrational states v and two different Jacobi coordinate systems (RC and PC) using the potential energy surfaces SM, PS, and AY ($P_{v'} = \sum_{j'} P(v',j')$, $P_{tot} = \sum_{v'} P_{v'}$).

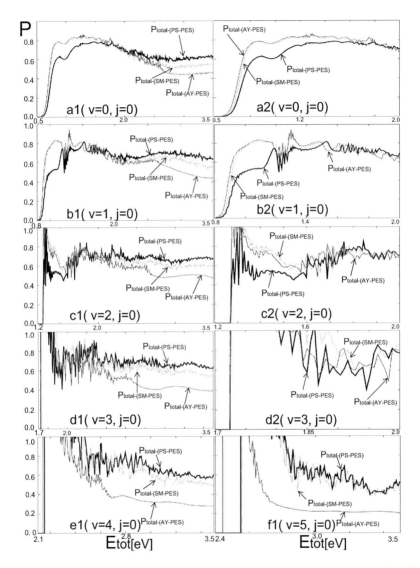

Figure 22: $H^- + H_2$ (v=0–5, j=0) (WP): Total reaction probabilities P as a function of total energy E_{tot} for the three different potential energy surfaces (PS-PES, SM-PES, and AY-PES) using product coordinates (PC).

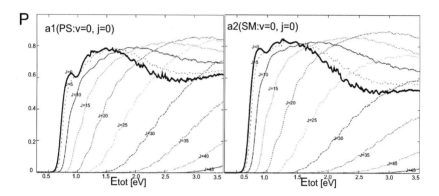

Figure 23: $H^- + H_2(v=0, j=0)$ (WP): Total reaction probabilities P (using reactant coordinates (RC)) as a function of total energy E_{tot} for different total angular momenta $J = 0$, 5, 10, 15, 20, 25, 30, 35, 40, 45.

angular momenta $J = 0$, 5, 10, 15, 20, 25, 30, 35, 40, 45 using reactants coordinates (RC) for PS-PES (left panel $a1$) and SM-PES (right panel $a2$). The total energies are in the range of 0.3–3.5 eV. From Fig. 23, it can be seen that the reaction barrier for each total angular momentum J is the same, whether investigated on PS-PES or on SM-PES. For low total angular momenta ($J \leq 20$), the curves for the reaction probabilities look similar for PS-PES and SM-PES. For $J = 0$, a maximum is reached for $E_{tot} = 1.2$–1.5 eV. For the same J values, the maxima of the total reaction probabilities are higher for SM-PES than for PS-PES. For $J > 20$ and higher total energies, the total reaction probabilities for PS-PES are higher than for SM-PES.

Fig. 24 shows the reaction cross sections σ calculated with the wave packet (WP) method (J up to 50) and quasi-classical trajectories (QCT), using the code CTAMYM (CT) for two different potential energy surfaces (PS-PES and SM-PES) for the collision $H^- + H_2(v=0, j=0)$. It can be seen from the lines a, c, e, and f in Fig. 24 that the results using WPs exhibit the same threshold for both PS-PES and SM-PES. When the total energy E_{tot} is larger than the reaction barrier, the total reaction cross section σ increases with an increase in the total energy. For SM-PES and $E_{tot} \simeq 1.8$ eV, σ reaches its maximum at $3.5*10^{-16}$ cm^2 using WPs. For PS-PES and WPs, σ is monotonically increases with E_{tot} in the range of 0.3–3.0 eV. Jaquet and Heinen [71] have calculated the total reaction cross section σ (J up to 20) using the J-shifting approximation for SM-PES. For SM-PES and E_{tot} between 0.6 and 0.9 eV our results are in quite good agreement with Jaquet and Heinen [71] (see lines a and e), but for total energies higher than 0.9 eV, our results are significantly lower than those calculated by Jaquet and Heinen [71]. Panda et al. [150] (see line f) have calculated the total reaction cross section σ using centrifugal sudden approximation for PS-PES. Using PS-PES, the results of Panda et al. [150] (see line f) is lower than our results (see line c) for the neglegence of the Coriolis

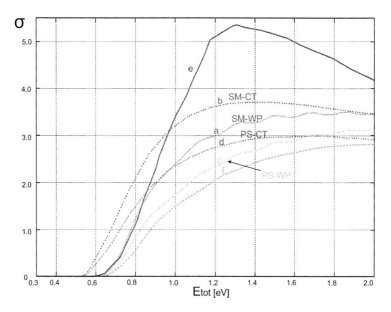

Figure 24: Reaction cross sections σ [in 10^{-16} cm²] for H⁻ + H₂(v=0, j=0): Comparison of reaction cross sections using quantum mechanical wave packets and quasi-classical trajectories (CTAMYM) for different potential energy surfaces (SM and PS). a: σ based on wave packets for SM-PES($J = 0$–50). b: σ calculated with CTAMYM for SM-PES. c: σ based on wave packets for PS-PES($J = 0$–50). d: σ calculated with CTAMYM for PS-PES. e: σ calculated by Jaquet and Heinen, $J = 0$–20 with J-shift [71] for SM-PES. f: σ calculated by Panda and Sathyamurthy [150] for PS-PES.

coupling. At low energies and using the same potential energy surface, the total reaction cross sections σ calculated using QCT (see lines b and d) are larger than those calculated using wavepackets (see lines a and c). For $E_{tot} > 1.6$ eV, the results using QCT are lower than those using WPs. Li and Wang [151] got the same characterics. The similar characterics are found by Jambrina et al. [152] for the H⁺ + D₂ reaction system. They found that as J increases, the QCT reaction probabilities decrease more rapidly with total energy than do those obtained using the WP. For the case of using the same method, σ for SM-PES is slightly higher than for PS-PES. This is because the reaction barrier for PS-PES ($E_{bar} = 0.469$ eV) is slightly higher than that for SM-PES ($E_{bar} = 0.465$ eV).

5.3 $H^- + D_2(v,j)$

5.3.1 Quasi-classical trajectory calculations

The $H^- + D_2(v=0-5, j=0-1) \to HD(v', j') + D^-$ reaction was studied with quasi-classical trajectories (QCT), using the code CTAMYM for the potential energy surfaces of Stärck and Meyer (SM-PES) [50] and Panda and Sathyamurthy (PS-PES) [53].

Initial parameter determinations:
1) The maximum impact parameter b_{max} was calculated in the same way as in the case of $H^- + H_2(v,j)$. The results are listed in Tab. 17.

Table 17: The impact parameters b_{max} for $H^- + D_2(v=0-5, j=0-1)$.

v	$v = 0$	$v = 1$	$v = 2$	$v = 3$	$v = 4$	$v = 5$
b_{max} [Å]	2.0	2.2	10.2	10.3	10.6	10.8

2) We investigated the initial internal energies, turning points, and vibrational half-periods for D_2 using SM-PES and PS-PES. The results are shown in Tab. 18 and Tab. 19.

Table 18: Internal energies, turning points, and vibrational half-periods of D_2 for different vibrational (v) and rotational (j) states (using SM-PES).

v	j	$E_{v,j}[\text{cm}^{-1}]$	r_- [Å]	r_+ [Å]	$\tau[fs]$
0	0	1538.832	0.649029	0.85892	0.5483902
0	1	1598.489	0.649568	0.85953	0.5489402
1	0	4522.894	0.593099	0.96191	0.5695584
1	1	4580.505	0.593618	0.96256	0.5699135
2	0	7395.162	0.559756	1.04352	0.5920615
2	1	7450.742	0.560263	1.04420	0.5925734
3	0	10154.712	0.535462	1.11750	0.6168351
3	1	10208.266	0.535960	1.11821	0.6173034
4	0	12800.540	0.516340	1.18797	0.6440391
4	1	12852.066	0.516832	1.18871	0.6446697
5	0	15331.599	0.500647	1.25691	0.6741609
5	1	15381.089	0.501134	1.25770	0.6746304

Reaction probabilities:
For the $H^- + D_2(v=0-5, j=0-1) \to HD(v', j') + D^-$ reaction the energy dependence of the total and state-to-state reaction probabilities (P_{tot}, $P_{v'}$) were calculated using the potential energy surfaces SM and PS (see Fig. 25 and 26). All the left panels in these figures are

Table 19: Internal energies, turning points, and vibrational half-periods of D_2 for different vibrational (v) and rotational (j) states (using PS-PES).

v	j	$E_{v,j}[\text{cm}^{-1}]$	r_- [Å]	r_+ [Å]	$\tau[fs]$
0	0	1544.245	0.650129	0.859522	0.5465750
0	1	1603.748	0.650661	0.860128	0.5470078
1	0	4533.322	0.594471	0.963108	0.5695768
1	1	4590.686	0.594985	0.963753	0.5697452
2	0	7402.132	0.561286	1.04532	0.5932492
2	1	7457.410	0.561789	1.04600	0.5937764
3	0	10153.783	0.537057	1.11982	0.6189461
3	1	10207.010	0.537553	1.12053	0.6195836
4	0	12789.895	0.517917	1.19068	0.6467493
4	1	12841.086	0.518409	1.19142	0.6472422
5	0	15311.739	0.502129	1.25984	0.6769037
5	1	15360.888	0.502620	1.26062	0.6774299

the reaction probabilities P using PS-PES, and the right panels are the reaction probabilities P using SM-PES. Some important parts are enlarged and plotted in the same panel. The maximum impact parameters b_{max} [Å] used in the calculations are shown as well. For $D_2(v=0)$ (see panels $a1$ ($j = 0$ using PS-PES), $a2$ ($j = 0$ using SM-PES), $b1$ ($j = 1$ using PS-PES), and $b2$ ($j = 1$ using SM-PES) in Fig. 25) the reaction barrier is $E_{bar} = 0.3$ eV. For collision energies $E_{coll} > E_{bar}$, the total reaction probability P_{tot} steeply increase with an increase in E_{coll}. For $D_2(v=0, j=0)$ the total reaction probability P_{tot} reaches its maximum at $E_{coll} = 1.5$ eV (PS-PES) and $E_{coll} = 1.3$ eV (SM-PES). For $D_2(v=0, j=1)$, i.e., rotationally excited, the maximum is reached at $E_{coll} = 1.6/1.5$ eV (PS/SM). The maximum value for the reaction probability is lower for the $D_2(v=0, j=1)$ reactant compared to the $D_2(v=0, j=0)$ reactant. These results mean that the reaction for the $D_2(v=0, j=0)$ reactant is more favored than for the $D_2(v=0, j=1)$ reactant. At low collision energies, the total reaction probabilities for PS-PES are lower than those for SM-PES.

The state-to-state reaction probabilities $P_{v'}$ for the $H^- + D_2(v=0, j=0-1)$ reaction have similar features as in the reaction $H^- + H_2(v=0, j=0-1)$. At low collision energies, the highest reaction probabilities are found for the HD product with final vibrational state $v'=0$. The HD products with the vibrational state $v'=1$ will be found only for $E_{coll} > 0.5$ eV. In order to get HD products with the vibrational state $v'=2$ the corresponding $E_{coll} > 1.0$ eV is needed.

The features of the total reaction probabilities P_{tot} for $H^- + D_2(v=1, j=0-1) \rightarrow HD(v', j')$ + D^- reaction are similar to the $H^- + D_2(v=0, j=0-1) \rightarrow HD(v', j') + D^-$ reaction. At the same collision energies E_{coll}, P_{tot} for the $D_2(j=1)$ reactant are lower than that for the $D_2(j=0)$ reactant. Total reaction probabilities P_{tot}, using PS-PES, are lower than those using SM-PES. Panel $c1$ in Fig. 25 shows the state-to-state reaction probabilities $P_{v'}$ for the $H^- + D_2(v=1,$

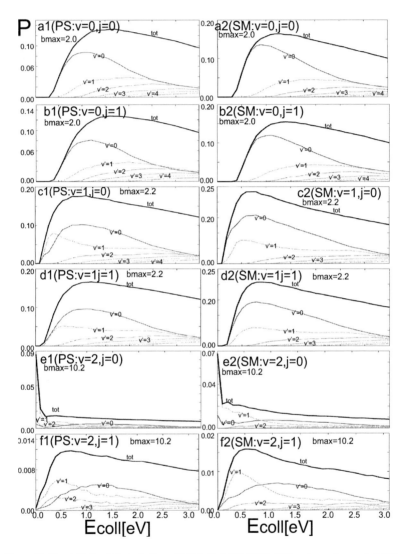

Figure 25: $H^- + D_2(v{=}0{-}2, j{=}0{-}1) \rightarrow HD(v', j') + D^-$ (CT): Different reaction probabilities P (P_{tot}, $P_{v'}$: see abbreviation *tot* or $v' = \cdot\cdot$) as a function of collision energy E_{coll} for various initial rovibrational states v, j and different impact parameters b_{max} using the potential energy surfaces SM and PS ($P_{v'} = \sum_{j'} P(v', j')$, $P_{tot} = \sum_{v'} P_{v'}$).

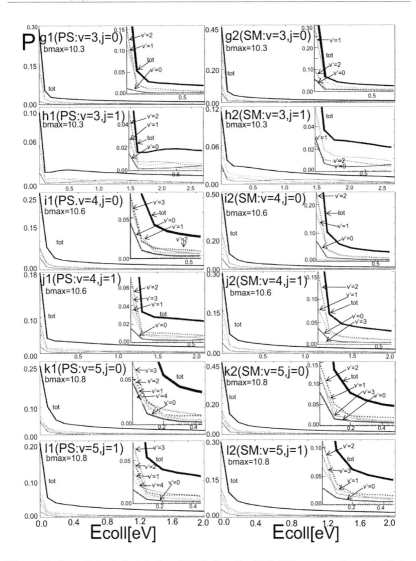

Figure 26: H⁻ + D₂(v=3–5, j=0–1) → HD(v', j') + D⁻ (CT): Different reaction probabilities P (P_{tot}, $P_{v'}$: see abbreviation *tot* or $v' = \cdots$) as a function of collision energy E_{coll} for various initial rovibrational states v, j and different impact parameters b_{max} using the potential energy surfaces SM and PS ($P_{v'} = \sum_{j'} P(v', j')$, $P_{tot} = \sum_{v'} P_{v'}$).

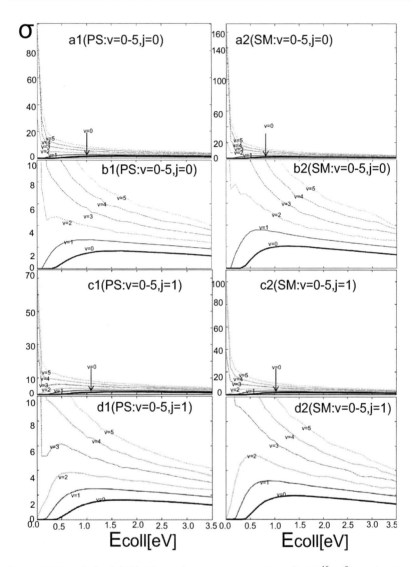

Figure 27: $H^- + D_2(v, j)$ (CT): The total reaction cross section σ (in 10^{-16} cm^2) as a function of collision energy E_{coll} using PS-PES and SM-PES. Plot $b1(b2,\ d1,\ d2)$ is an enlargement of plot $a1(a2,\ c1,\ c2)$.

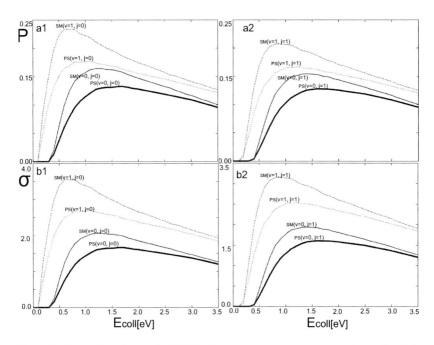

Figure 28: $H^- + D_2(v{=}0{-}1,\ j{=}0{-}1)$ (CT): Comparison of reaction cross sections σ (in 10^{-16} cm^2) and reaction probabilities P as a function of collision energy E_{coll} for different potential energy surfaces (SM, PS).

$j{=}0) \to$ HD(v',j') + D$^-$ reaction using PS-PES. In this case, HD($v'{=}1$) is produced at low collision energies ($E_{coll} \leq 0.5$ eV), but at higher collision energies $E_{coll} > 0.5$ eV, HD($v'{=}0$) is strongly favored. For SM-PES (see panel $c2$ in Fig. 25) HD($v'{=}0$) is the main product over the entire energy range. HD($v'{=}1$) is a minor product at low collision energies. The panels $d1$ and $d2$ in Fig. 25 show that for the $H^- + D_2(v{=}1,\ j{=}1) \to$ HD(v',j') + D$^-$ reaction, the main product is HD($v'{=}0$). Similar feature are seen for SM-PES and PS-PES.

For the $D_2(v{=}2)$ reactant (see panels $e1$, $e2$, $f1$ and $f2$ in Fig. 25) and low collision energies ($E_{coll} \leq 0.8$ eV) the main product is HD($v'{=}1$), but at higher collision energies $E_{coll} = 0.8\text{--}3.0$ eV the main product is HD($v'{=}0$).

Fig. 26 shows the total and state-to-state reaction probabilities (P_{tot}, $P_{v'}$) for reactants $D_2(v{=}3{-}5)$. In these cases P_{tot} is high at low collision energies ($E_{coll} = 0.01$ eV), which decreases with increase of collision energy. For $E_{coll} > 0.3$ eV, P_{tot} decreases slowly with an increase in collision energy. For PS-PES and the state-to-state reaction probabilities for the $H^- + D_2(v{=}3{-}4,\ j{=}0{-}1) \to$ HD(v',j') + D$^-$ reaction, the most favored vibrational state for the HD product is $v' = 2$. For $H^- + D_2(v{=}5,\ j{=}0{-}1) \to$ HD(v',j') + D$^-$ and low collision energies, the most important vibrational state of the HD product is $v' = 3$. For SM-PES the $H^- + D_2(v{=}3, j{=}0{-}1) \rightarrow HD(v', j') + D^-$ reaction leads to the most favored vibrational states of the HD product with $v' = 1$. For the $D_2(v{=}4{-}5, j{=}0{-}1)$ reactants and low collision energies, the main vibrational state of the HD product is $v' = 2$.

Table 20: $H^- + D_2(v, j{=}0)$ (CT): Energy dependence of the total reaction cross sections σ [10^{-16} cm^2] using PS-PES for different initial vibrational states v of D_2.

$E[eV]$	$v = 0$	$v = 1$	$v = 2$	$v = 3$	$v = 4$	$v = 5$
0.01	0.00000	0.00000	28.2759	95.8713	94.0293	93.1661
0.10	0.00000	0.01490	7.44567	17.8177	27.8261	38.0359
0.20	0.00000	0.81227	4.49421	9.48881	14.6420	20.2015
0.30	0.00000	1.51050	4.70993	8.45894	11.9310	15.5625
0.40	0.16022	1.99676	4.79164	7.93900	10.7132	13.8292
0.50	0.46169	2.30056	4.64783	7.31242	9.88723	12.6420
0.60	0.75788	2.50736	4.46152	6.84914	9.22008	11.4181
0.70	0.99425	2.56939	4.42230	6.42919	8.44351	10.6596
0.80	1.19644	2.67218	4.20004	6.07924	7.99874	9.95605
0.90	1.33769	2.66960	4.10198	5.77928	7.64928	9.27448
1.00	1.44073	2.67492	3.99412	5.64929	7.13038	8.73582
1.10	1.52744	2.69925	3.89280	5.20935	6.83034	8.30343
1.20	1.60498	2.64937	3.82416	5.14935	6.57266	7.98829
1.30	1.63149	2.61607	3.69342	4.97271	6.45265	7.47895
1.40	1.64682	2.58749	3.65420	4.83273	5.91963	7.29939
1.50	1.64821	2.56270	3.53653	4.62275	5.80315	7.00258
1.60	1.66366	2.53716	3.37964	4.47944	5.74667	6.85601
1.70	1.66957	2.47603	3.34042	4.37279	5.55605	6.63614
1.80	1.65022	2.44319	3.30447	4.38612	5.36544	6.50789
1.90	1.62383	2.41080	3.22602	4.25947	5.21718	6.21108
2.00	1.61063	2.36731	3.20968	4.17614	5.16423	5.92160
2.10	1.58663	2.33766	3.19987	4.01283	4.98421	5.78968
2.50	1.49678	2.21648	2.85668	3.66621	4.42649	5.35729
3.00	1.37262	2.03629	2.65730	3.37624	4.13351	4.76000
3.50	1.20323	1.84562	2.42851	3.05628	3.58284	4.01613

Reaction cross sections:

First, the reaction cross sections σ for the reaction $H^- + D_2(v{=}0{-}5, j{=}0{-}1) \rightarrow HD(v', j') + D^-$ were calculated with quasi-classical trajectories using SM-PES and PS-PES. The results are shown in Fig. 27 and Tabs. 20–23. The collision energies E_{coll} were chosen in the range of 0.01–3.5 eV. The left panels in Fig. 27 show the reaction cross sections σ for different initial

Table 21: $H^- + D_2(v, j=1)$ (CT): Energy dependence of the total reaction cross sections σ [in 10^{-16} cm^2] using PS-PES for different initial vibrational states v of D_2.

$E[eV]$	$v = 0$	$v = 1$	$v = 2$	$v = 3$	$v = 4$	$v = 5$
0.01	0.00000	0.00000	0.15689	31.0061	62.8709	74.5805
0.10	0.00000	0.00000	1.04919	5.14269	11.7333	19.8534
0.20	0.00000	0.13958	1.69963	5.17935	9.73192	14.5987
0.30	0.00000	0.67770	2.77170	5.75261	9.44600	13.1293
0.40	0.04411	1.26615	3.46789	6.08257	9.05065	12.2316
0.50	0.25661	1.71242	3.72937	6.15590	8.58117	11.0480
0.60	0.50743	2.03462	3.81109	5.99258	8.19288	10.4250
0.70	0.76605	2.20371	3.84377	5.72595	7.73400	9.87543
0.80	0.99237	2.37355	3.78821	5.61596	7.43043	9.19753
0.90	1.17119	2.44562	3.77186	5.40932	7.13391	8.86774
1.00	1.29346	2.50781	3.66727	5.20601	6.67856	8.58192
1.10	1.40505	2.52165	3.57902	4.99271	6.63973	8.12387
1.20	1.47579	2.53199	3.52346	4.96938	6.43853	7.87103
1.30	1.53724	2.51359	3.48423	4.84273	6.16319	7.33970
1.40	1.58010	2.50401	3.50711	4.65275	5.96199	7.24443
1.50	1.59417	2.48500	3.40906	4.53943	5.67254	7.01357
1.60	1.60661	2.45642	3.31427	4.40611	5.57723	6.79371
1.70	1.60234	2.42509	3.23910	4.21614	5.39368	6.58118
1.80	1.60058	2.38845	3.21622	4.23614	5.32308	6.35765
1.90	1.59015	2.37324	3.15412	4.14948	5.11835	6.17077
2.00	1.57884	2.32565	3.14431	4.13615	4.95950	6.00588
2.10	1.55961	2.30984	3.14758	4.01950	4.88890	5.81167
2.50	1.49464	2.19337	2.89917	3.74953	4.37001	5.20705
3.00	1.36169	2.01714	2.58213	3.33292	4.06291	4.58044
3.50	1.19896	1.83437	2.42524	3.01296	3.54754	4.16271

vibrational states using PS-PES. The panels $b1$ and $d1$ are the enlargement of panels $a1$ and $c1$. The right panels show the total reaction cross sections σ using SM-PES; panels $b2$ and $d2$ are the enlargement of the panels $a2$ and $c2$.

Fig. 27 demonstrates that for the $H^- + D_2(v=0, j=0) \rightarrow HD(v', j') + D^-$ reaction the total reaction cross section increases dramatically when the collision energy increases up to the threshold. The reaction cross section reaches its maximum at collision energies $E_{coll} = 1.7/1.3$ eV using PS-PES/SM-PES. After the maximum the reaction cross section decreases slowly with an increase in E_{coll}. The results are quite similar for PS-PES and SM-PES; at low collision energies the reaction cross sections, using SM-PES, are much higher than those using PS-PES. For the $D_2(v=0, j=1)$ reactant, as shown in Fig. 27 and Tabs. 21 and 23,

73

Table 22: $H^- + D_2(v, j{=}0)$ (CT): Energy dependence of the total reaction cross sections σ [in 10^{-16} cm^2] using SM-PES for different initial vibrational states v of D_2.

$E[eV]$	$v = 0$	$v = 1$	$v = 2$	$v = 3$	$v = 4$	$v = 5$
0.01	0.00000	0.00000	22.2291	152.340	172.460	170.289
0.10	0.00000	0.00608	7.24956	22.3971	36.6191	48.6296
0.20	0.00000	1.35510	7.77252	14.5781	21.9277	28.6222
0.30	0.00000	2.26939	6.83446	12.3017	16.9787	22.0117
0.40	0.24756	2.98830	6.86715	10.8853	14.7620	18.4463
0.50	0.77673	3.38242	6.55664	9.99875	13.0676	16.6031
0.60	1.24319	3.54968	6.03041	9.02887	12.0898	14.7087
0.70	1.56904	3.55150	5.82122	8.30896	10.7979	13.5837
0.80	1.78040	3.55348	5.41919	7.65237	10.0496	12.3342
0.90	1.90858	3.43838	5.27211	7.20576	9.22008	11.3558
1.00	1.97682	3.38516	4.94199	6.72582	8.57764	10.5899
1.10	2.04229	3.32738	4.68705	6.33254	8.17170	9.97071
1.20	2.06352	3.21516	4.43210	6.04591	7.74106	9.34777
1.30	2.06667	3.13351	4.33732	5.79927	7.40572	8.75781
1.40	2.04266	3.05277	4.09872	5.61263	6.74563	8.40236
1.50	2.03676	2.98404	4.03335	5.21601	6.60796	8.13853
1.60	2.01389	2.91805	3.87972	5.02937	6.44206	7.76843
1.70	1.98674	2.83229	3.69996	4.93272	6.24085	7.40932
1.80	1.93195	2.77254	3.67381	4.81940	6.01141	7.21878
1.90	1.89149	2.71673	3.57902	4.61942	5.72196	6.96594
2.00	1.85241	2.65378	3.56268	4.48277	5.60547	6.56652
2.10	1.81383	2.60756	3.49731	4.32946	5.41839	6.32834
2.50	1.66555	2.39696	3.04625	3.90618	4.79360	5.70540
3.00	1.47001	2.15368	2.75863	3.56289	4.33118	5.00551
3.50	1.26405	1.92742	2.54290	3.20960	3.61814	3.86223

the reaction cross section increases for $E_{coll} > 0.4$ eV. The maximum of the cross section is reached at $E_{coll} = 1.6$ and 1.4 eV using PS-PES and SM-PES. The maxima of these reaction cross sections are lower than those for the $H^- + D_2(v{=}0,\ j{=}0) \rightarrow HD(v', j') + D^-$ reaction. For the $D_2(v{=}1)$ reactant the cross section starts to increase at $E_{coll} = 0.1$ eV and at $E_{coll} = 0.2$ eV for the $D_2(j{=}0)$ and $D_2(j{=}1)$ reactants. The maximum of the reaction cross section, using PS-PES, is reached at $E_{coll} = 1.10$ and 1.20 eV for the $D_2(j{=}0)$ and $D_2(j{=}1)$ reactants. When using SM-PES the maximum of the reaction cross section is reached at $E_{coll} = 0.8$ and 0.9 eV for the $D_2(j{=}0)$ and $D_2(j{=}1)$ reactants. All the collision energies corresponding to these maximum values are lower than those for the $D_2(v{=}0)$ reactant.

For the $D_2(v{=}2)$ reactant, the reaction cross sections are much different for different initial

Table 23: $H^- + D_2(v, j{=}1)$ (CT): Energy dependence of the total reaction cross sections σ [in 10^{-16} cm^2] using SM-PES for different initial vibrational states v of D_2.

$E[eV]$	$v = 0$	$v = 1$	$v = 2$	$v = 3$	$v = 4$	$v = 5$
0.01	0.00000	0.00000	0.00000	37.6219	100.587	106.621
0.10	0.00000	0.00000	1.17340	12.3284	23.9468	34.9982
0.20	0.00000	0.08652	3.19007	9.34216	16.3716	22.7409
0.30	0.00000	1.06635	4.52689	9.14885	14.1584	19.0546
0.40	0.03732	1.97030	5.04658	8.74224	12.7146	16.9549
0.50	0.42097	2.55647	5.20674	8.33562	11.6204	15.1521
0.60	0.82222	2.89844	5.19367	7.88901	10.5296	13.6277
0.70	1.19305	3.08044	5.14137	7.37908	9.85899	12.4331
0.80	1.48484	3.14948	4.90604	7.07245	9.12477	11.5390
0.90	1.64469	3.15738	4.75569	6.60917	8.67295	10.6706
1.00	1.77073	3.13914	4.53996	6.40586	8.17170	10.1502
1.10	1.86636	3.12515	4.37000	5.99258	7.81871	9.56030
1.20	1.89878	3.05247	4.21638	5.83260	7.49749	9.13890
1.30	1.93472	2.97842	4.10525	5.59930	7.17627	8.52329
1.40	1.94251	2.94633	4.03008	5.37266	6.86211	8.25579
1.50	1.93359	2.90087	3.87646	5.24601	6.51971	8.05059
1.60	1.92379	2.83883	3.79474	5.01604	6.31145	7.73545
1.70	1.88747	2.75687	3.64766	4.80606	6.11378	7.42398
1.80	1.87578	2.70609	3.57248	4.73607	5.96199	7.15648
1.90	1.83255	2.66245	3.49077	4.56609	5.67960	6.86333
2.00	1.80704	2.59798	3.50711	4.45278	5.45722	6.63614
2.10	1.77224	2.56331	3.47770	4.36612	5.31955	6.26238
2.50	1.64330	2.36914	3.13777	3.99950	4.69829	5.70906
3.00	1.46536	2.14136	0.00219	3.48290	4.28176	4.78931
3.50	1.25827	1.90963	2.52983	3.09961	3.57578	4.07843

rotational states $j = 0$ and $j = 1$. For the $D_2(v{=}2, j{=}0)$ reactant and low collision energies the reaction cross sections are high. The reaction cross section decreases with an increase in collision energy. But in the case of the $D_2(j{=}1)$ reactant and low collision energies, the reaction cross sections are low.

For the $D_2(v{=}3,4,5)$ reactants the reaction cross sections look similar; the reaction cross sections are high at low collision energies $E_{coll} = 0.01$ eV. With an increase in E_{coll} from 0.01 to 0.5 eV the reaction cross section decreases rapidly. For $E_{coll} > 0.5$ eV the reaction cross section decreases slowly with an increase in collision energy.

In order to compare the different reaction cross sections for PS-PES and SM-PES for the $H^- + D_2(v{=}0{-}1, j{=}0{-}1)$ reaction, the energy dependence of the reaction cross sections and reaction probabilities are shown together in the same figure (see Fig. 28). For the D_2 reactant, starting with the same initial rovibrational states, the reaction cross sections using SM-PES are higher than those using PS-PES. For the D_2 reactant, starting with the same initial vibrational states, the reaction cross sections for the $D_2(j=0)$ reactant are higher than those for the $D_2(j=1)$ reactant. At higher collision energies, the differences in the reaction cross section become small.

5.3.2 Wave packet calculations for $H^- + D_2(v,j) \rightarrow HD(v',j') + D^-$

Reaction probabilities P for the ion-molecule collisions $H^- + D_2(v,j)$ for total angular momenta $J = 0$ and $J \neq 0$ were investigated by time-dependent wave packets (WPs) using the real wave packet code of S. Gray [149]. The parameters used in the calculations are the same as those for $H^- + H_2(v,j) \rightarrow H_2(v',j') + H^-$, as shown in Tab. 16.

Time-dependent quantum dynamics of $H^- + D_2(v=0–1, j=0) \rightarrow HD(v',j') + D^-$ *for total angular momentum J = 0*:
The ion-molecule reaction $H^- + D_2(v=0–1, j=0) \rightarrow HD(v',j') + D^-$ for total angular momentum $J = 0$ was investigated for the dynamics of reactive scattering processes using WPs for PS-PES, SM-PES, and AY-PES. In order to make sure that the results for the reaction probabilities are converged, firstly, reactant Jacobi coordinates (RC) are used to calculate the total reaction probabilities and then these are compared with those calculated using product Jacobi coordinates (PC) (see left panels in Figs. 29). These panels show that the reaction probabilities, which are calculated using PC-coordinates and RC-coordinates, are the same if the D_2 reactant starts with the initial rovibrational states $v=0–1$, $j=0$. The total and state-to-state reaction probabilities (P_{tot}, $P_{v'}$) are determined by using product Jacobi coordinates (PC) for PS-PES, SM-PES, and AY-PES, as shown in the right panels of Fig. 29. For the $D_2(v=0, j=0)$ reactant, the total reaction probabilities P_{tot} for the three PES are increased immediately at 0.5 eV up to 0.7 eV, with a maximum value of ~40%, ~55%, and ~55% using PS-PES, SM-PES, and AY-PES, respectively; this is followed by a slow increase up to 1.4 eV with a magnitude of ~60%, ~70%, and ~70%. For $E_{tot} = 1.5$–2.7 eV, the total reaction probabilities decrease slowly with an increase in total energy. At low energies, the largest reaction probability is found for the $HD(v'=0)$ product. The other vibrational states become populated with an increase in E_{tot}. For the $D_2(v=1, j=0)$ reactant the reaction probability characteristics are the same as those for the $D_2(v=0, j=0)$ reactant, i.e., at low E_{tot}, there is a steep increase in the total reaction probabilities up to the maximum. Thereafter, the total reaction probabilities decrease with an increase in E_{tot}. For $E_{tot} = 0.5$–0.8 eV, the main products are $HD(v'=0)$. When the total energies are in the range of 0.8–1.2 eV, the main product HD is found with the vibrational state of $v' = 1$. At higher total energies $E_{tot} > 1.2$ eV, $HD(v'=0)$ is the main product again. HD products with the higher final vibrational state will be found one by one with an increase in E_{tot}.
In order to show the differences in the total reaction probabilities using the three potential

76

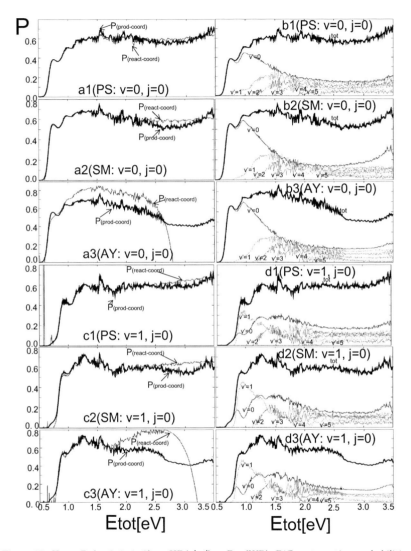

Figure 29: $H^- + D_2(v{=}0{-}1,\ j{=}0) \rightarrow HD(v', j') + D^-$ (WP): Different reaction probabilities P (P_{tot}, $P_{v'}$: see abbreviation tot or $v' = \cdots$) as a function of total energy E_{tot} for various initial vibrational states v and two different Jacobi coordinate systems (RC and PC) using the potential energy surfaces SM, PS, and AY ($P_{v'} = \sum_{j'} P(v', j')$, $P_{tot} = \sum_{v'} P_{v'}$).

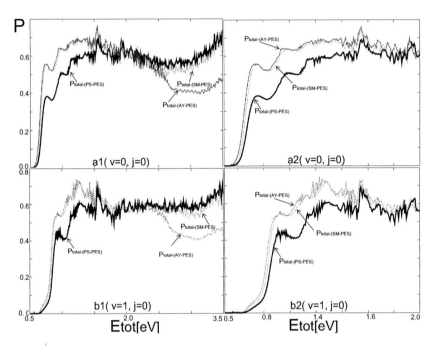

P

0.6

0.4

$P_{total-(PS-PES)}$ $P_{total-(SM-PES)}$

$P_{total-(AY-PES)}$

0.2

a1(v=0, j=0)

$P_{total-(AY-PES)}$

$P_{total-(SM-PES)}$

$P_{total-(PS-PES)}$

a2(v=0, j=0)

0.8

0.6

$P_{total-(SM-PES)}$

$P_{total-(PS-PES)}$ $P_{total-(AY-PES)}$

0.4

0.2

b1(v=1, j=0)

$P_{total-(AY-PES)}$

$P_{total-(SM-PES)}$

$P_{total-(PS-PES)}$

b2(v=1, j=0)

0.5   2.0   3.5 0.5   0.8   1.4   1.6   2.0

Etot[eV]   Etot[eV]

Figure 30: H^- + $D_2(v{=}0{-}1,\ j{=}0)$ (WP): Comparison of total reaction probabilities P as a function of total energy E_{tot} for three different potential energy surfaces (PS-PES, SM-PES, and AY-PES) using product coordinates (PC).

energy surfaces (SM, PS, and AY), results of the reaction probabilities are plotted in the same panel in Fig. 30. For the reaction H^- + $D_2(v{=}0{-}1,\ j{=}0) \rightarrow HD(v',j')$ + D^- and $E_{tot} < 2.0$ eV, the total reaction probabilities P_{tot} using SM-PES and AY-PES are similar. Using PS-PES, P_{tot} is much lower than for SM-PES and AY-PES. For $E_{tot} > 2.5$ eV, the reaction probabilities using AY-PES are much lower than those using SM-PES and PS-PES.

Time-dependent quantum dynamics of H^- + $D_2(v{=}0,\ j{=}0) \rightarrow HD(v',j')$ + D^- *for total angular momenta* $J \neq 0$.
Fig. 31 shows the total reaction probabilities using the wave packet (WP) program for H^- + $D_2(v{=}0,\ j{=}0)$ and total angular momenta $J = 0, 5, 10, 15, 20, 25, 30, 40$ for PS-PES and SM-PES. The trend of the curves and the threshold energy in the calculations are similar. The reaction probabilities are different using PS-PES and SM-PES. For each total angular momentum the maxima of the reaction probability using SM-PES are much higher than those using PS-PES.

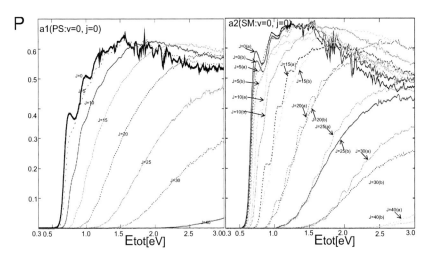

Figure 31: $H^- + D_2(v=0, j=0)$ (WP): Total reaction probabilities P (using reactant coordinates (RC)) as a function of total energy E_{tot} for different total angular momenta $J = 0, 5, 10, 15, 20, 25, 30, 40$. In the right panel: (a) are Morari and Jaquet (2005) [72] results, (b) are present results.

Fig. 32 shows the total reaction cross sections σ [10^{-16} cm^2] for PS-PES and SM-PES compared with the experimental results by Zimmer and Linder [65], Haufler et al. [67], and Huq et al. [63]. In Fig. 32 line a is cross section σ using SM-PES based on wave packets (WPs) ($J = 0$–50). Line b is σ using SM-PES for quasi-classical trajectories (QCT) using the code CTAMYM. It can be seen that when E_{tot} are in the range of 0.3–2.0 eV, σ, shown by line b, is slightly higher than in the case of line a. For E_{tot} in the range of 0.5–0.7 eV, the results of σ, calculated with QCT, are in good agreement with the experimental results by Haufler et al. [67]. At higher energies ($1.2 \leq E_{tot} \leq 2.0$ eV), the results of σ using QCT are in good agreement with the experimental cross sections of Zimmer and Linder [65]. As shown by line a, the results calculated with WPs in the present work are closer to the experimental results by Haufler et al. [67] than others. For $E_{tot} \geq 1.2$ eV, the results in the present work are in good agreement with the results of Zimmer and Linder [65]. The values showed by line a are slightly lower than the results of Morari and Jaquet [72] (see line c).

Line d shows σ based on wave packets (WPs) ($J = 0$–50) using PS-PES. Line e shows σ calculated with quasi-classical trajectories (QCT) using PS-PES. The characteristic trends of σ are similar to those using SM-PES, i.e., at low total energies ($E_{tot} < 1.5$ eV), σ calculated with QCTs is higher compared to WP-calculations. For $E_{tot} < 1.0$ eV, σ shown by line e, is in good agreement with the experimental results of Haufler et al. [67] (see line i). For $E_{tot} > 1.5$ eV, the results of σ using PS-PES are greater than the experimental results of Haufler

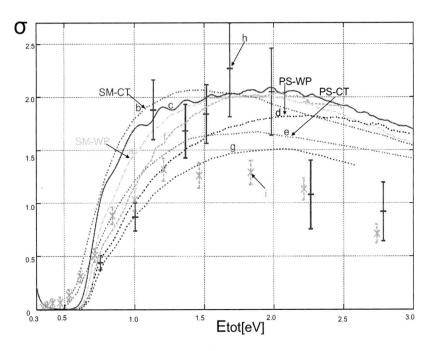

Figure 32: Reaction cross sections σ [in 10^{-16} cm²] for $H^- + D_2(v{=}0, j{=}0)$: Comparison of theoretical (quantum and classical) and experimental results. a: σ based on wave packets (WPs) ($J = 0$–50) for SM-PES. b: σ calculated with quasi-classical trajectories (QCT) using SM-PES. c: Morari and Jaquet (2005) [72] (σ based on WPs: $\Omega = 8$ ($J = 0$–60)) using SM-PES. d: σ based on WPs ($J = 0$–50) using PS-PES. e: σ calculated with QCTs using PS-PES. f: σ of Panda et al. [150] using PS-PES. g: σ of Yao et al. [153] using PS-PES. h: Zimmer and Linder (1995) [65]. i: Haufler et al. (1997) [67].

et al. [67]. As shown in Fig. 32, the results shown by line d are closer to the experimental results by Haufler et al. [67] than others, but σ based on WPs using PS-PES is smaller than the experiment results for $E_{tot} < 0.8$ eV. For $E_{tot} > 1.2$ eV, σ, shown by line d, is much higher than the experimental results. The reason that at higher energies the theoretical results are higher than the experimental results will be discussed in the $D^- + H_2$ chapter.

Fig. 32 shows that the present results of σ based on WPs using PS-PES are lower than the theoretical results of Panda et al. [150] (see line f) (they calculated the total reaction cross section without the Coriolis coupling). The present results of σ are higher than those of the theoretical work of Yao et al. [153] (see line g). A possible explanation for this finding is that in the present work the integral cross section was calculated up to $\Omega = J$, whereas Yao et al.

used a maximum value of $\Omega_{max} = 9$ (J =0-65).

5.4 $D^- + H_2(v, j)$

A detailed study of the proton exchange reaction $D^- + H_2(v, j) \rightarrow HD(v', j') + H^-$ on its ground potential energy surface was carried out using wave packets (S. Gray [149] program) and quasi-classical trajectories (QCTs) (CTAMYM program). The energy dependence of the total reaction probabilities P_{tot}, state-to-state reaction probabilities $P_{v'}$ and reaction cross sections σ for the $D^- + H_2(v, j) \rightarrow HD(v', j') + H^-$ reaction were calculated. The results in the present work will be presented in comparison with previous experiments.

5.4.1 Quasi-classical trajectory investigations

The $D^- + H_2(v=0-5, j=0-1) \rightarrow HD(v', j') + H^-$ was studied by QCTs using the potential energy surfaces SM and PS.

Initial parameter determinations:
The initial parameters for $D^- + H_2(v, j)$ were calculated in the same way as in the case of $H^- + H_2(v, j)$. The maximum impact parameter b_{max} is shown in Tab. 13. The internal energies, turning points, and vibration half-periods are shown in Tabs. 14 and 15.

Reaction probabilities:
The total and state-to-state reaction probabilities $(P_{tot}, P_{v'})$ for $D^- + H_2(v=0-5, j=0-1) \rightarrow HD(v', j') + H^-$ were calculated using PS-PES and SM-PES. The results for total and state-to-state reaction probabilities are shown in Figs. 33 and 34. All the left panels in these figures show the reaction probabilities using PS-PES, and the right panels show the reaction probabilities using SM-PES. Some important parts are enlarged in the same panel. The maximum impact parameters b_{max} [Å] used in the calculations are shown as well.

Panels $a1$, $a2$, $b1$, and $b2$ in Fig. 33 show the different reaction probabilities for the $H_2(v=0)$ reactant. It can be seen that these reactions have reaction barriers E_{bar}. For $E_{coll} > E_{bar}$ the total reaction probabilities steeply increase with an increase in collision energy E_{coll}. Beyond the maximum, the total reaction probabilities decrease slowly with an increase in E_{coll}. The maximum of the reaction probability using SM-PES is higher than that using PS-PES. It can be seen that the maximum of the total reaction probability for $D^- + H_2(v=0, j=0)$ is larger than that for $D^- + H_2(v=0, j=1)$. The state-to-state reaction probabilities show that for the $H_2(v=0)$ reactant and $E_{coll} = 0.3-3.5$ eV the main final vibrational state for HD products is $v' = 0$. With an increase in E_{coll} HD products with higher final vibrational states can be found. The total reaction probabilities for $D^- + H_2(v=1, j=0) \rightarrow HD(v', j') + H^-$ using PS-PES are similar to using SM-PES (see panels $c1$ and $c2$ in Fig. 33). For $E_{coll} = 0.01$ eV the total reaction probabilities are $\sim$40%. The total reaction probabilities increase with an increase in collision energy; beyond its maximum the total reaction probabilities decrease slowly with increase of E_{coll}. The differences between calculations using SM-PES and PS-PES are pronounced near the maximum of P_{tot} at $E_{coll} = 0.4/0.1$ eV for SM-PES/PS-PES. In the case of using PS-PES,

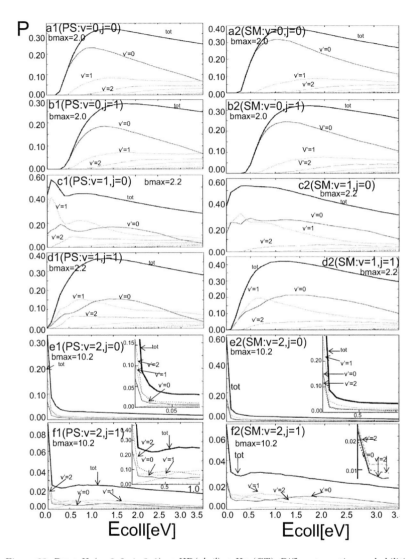

Figure 33: $D^- + H_2(v{=}0{-}2, j{=}0{-}1) \rightarrow HD(v',j') + H^-$ (CT): Different reaction probabilities P (P_{tot}, $P_{v'}$: see abbreviation tot or $v' = \cdot\cdot$) as a function of collision energy E_{coll} for various initial rovibrational states v,j and different impact parameters b_{max} using the potential energy surfaces SM and PS ($P_{v'} = \sum_{j'} P(v',j')$, $P_{tot} = \sum_{v'} P_{v'}$).

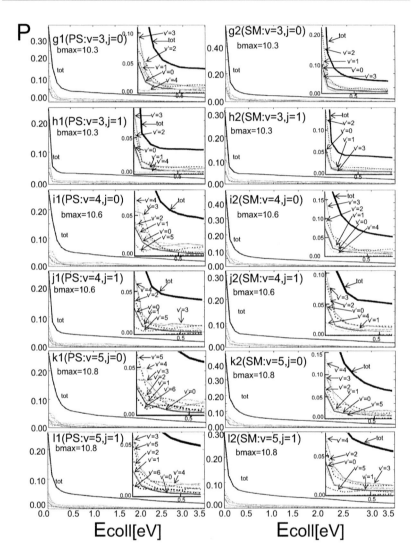

Figure 34: $D^- + H_2(v{=}3{-}5,\ j{=}0{-}1) \to HD(v',j') + H^-$ (CT): Different reaction probabilities P (P_{tot}, $P_{v'}$: see abbreviation tot or $v' = \cdots$) as a function of collision energy E_{coll} for various initial rovibrational states v,j and different impact parameters b_{max} using the potential energy surfaces SM and PS ($P_{v'} = \sum_{j'} P(v',j')$, $P_{tot} = \sum_{v'} P_{v'}$).

the state-to-state reaction probabilities show that for $E_{coll} < 1.3$ eV, the products are mainly HD with a final vibrational state $v' = 1$. For $1.3 < E_{coll} < 2.2$ eV, the HD(v'=0) product is strongly favored, and for even higher energy ($E_{coll} > 2.2$ eV) again the HD(v'=1) product predominates. For SM-PES and $E_{coll} < 0.9$ eV, the HD products mainly have a final vibrational state $v' = 1$; for $E_{coll} > 0.9$ eV the favored products are HD(v'=0). The characteristic of the total reaction probabilities for $D^- + H_2(v=1, j=1) \rightarrow HD(v', j') + H^-$ are similar to $D^- + H_2(v=0, j=1) \rightarrow HD(v', j') + H^-$. For $E_{coll} > E_{bar}$, the total reaction probabilities increase quickly with an increase in E_{coll}. The trend in the state-to-state reaction probabilities for the $H_2(v=1, j=1)$ reactant is similar to that for $D^- + H_2(v=1, j=0) \rightarrow HD(v', j') + H^-$.

In the case of $D^- + H_2(v=2-5, j=0-1)$, the trends in the total reaction probabilities are similar. The total reaction probabilities are higher at $E_{coll} = 0.01$ eV, then the total reaction probabilities steeply decrease with an increase in collision energy, and for higher collision energies $E_{coll} > 1.0$ eV, the reaction probabilities decreased slowly with an increase in E_{coll}. Using PS-PES and $E_{coll} = 0.01$ eV for $D^- + H_2(v=2-4, j=0-1) \rightarrow HD(v', j') + H^-$, the main products are HD(v'=v). For the $H_2(v=5, j=0)$ reactant the largest reaction probability is found for HD(v'=5). For the $H_2(v=5, j=1)$ reactant the main products are HD(v'=3, 4, 5). Using SM-PES and $E_{coll} = 0.01$ eV for $D^- + H_2(v=2-5, j=0-1) \rightarrow HD(v', j') + H^-$ the main final product is HD with a final vibrational state of v'=v−1.

Reactive cross sections:

Energy dependence of the reaction cross sections σ [10^{-16} cm^2] for the $D^- + H_2(v=0-5, j=0-1) \rightarrow HD(v', j') + H^-$ are calculated with Eq. (58) as shown in Fig. 35. The collision energies were chosen in the range of 0.01–3.5 eV. Left panels in Fig. 35 show σ for the H_2 reactant with different initial vibrational states using PS-PES. Panels $b1$ and $d1$ are the enlargements of $a1$ and $c1$. Right panels show σ using SM-PES, and panels $b2$ and $d2$ are the enlargements of $a2$ and $c2$.

Panels $a1$, $a2$, $b1$, and $b2$ in Fig. 35 demonstrate that, in the case of the $H_2(v=0, j=0)$ reactant, σ increases dramatically with an increase in E_{coll} near the threshold. For $E_{coll} \approx 1.3/1.4$ eV, σ reaches its maximum at 4.827 and 4.048*10^{-16} cm^2 using SM-PES/PS-PES. After these maxima, σ slowly decreases with an increase in E_{coll}. At low collision energy σ for SM-PES is much higher than in the case of PS-PES. In the case of the $H_2(v=0, j=1)$ reactant, as shown in panels $c1$, $c2$, $d1$, and $d2$ in Fig. 35, the reaction barrier is $E_{bar} = 0.5$ eV. For $E_{coll} > E_{bar}$, σ steeply increases with an increase in E_{coll}. The maxima of the cross sections are 4.030 and 3.550*10^{-16} cm^2 using SM-PES and PS-PES. These values are much lower than those for the $H_2(v=0, j=0)$ reactant.

The reaction cross sections for $D^- + H_2(v=1, j=0)$ are shown in panels $a1$, $a2$, $b1$, and $b2$ in Fig. 35. Using PS-PES σ shows a very sharp peak at $E_{coll} = 0.1$ eV with $\sigma = 8.530*10^{-16}$ cm^2. σ for SM-PES is different from the one for PS-PES, i.e., $\sigma \approx 6*10^{-16}$ cm^2 at $E_{coll} = 0.01$ eV. At $E_{coll} = 0.4$ eV, σ reaches its maximum ($\sigma_{max} = 8.197*10^{-16}$ cm^2). For $D^- + H_2(v=1, j=1)$ (see panels $d1$ and $d2$ in Fig. 35), the trends of the reaction cross sections are similar

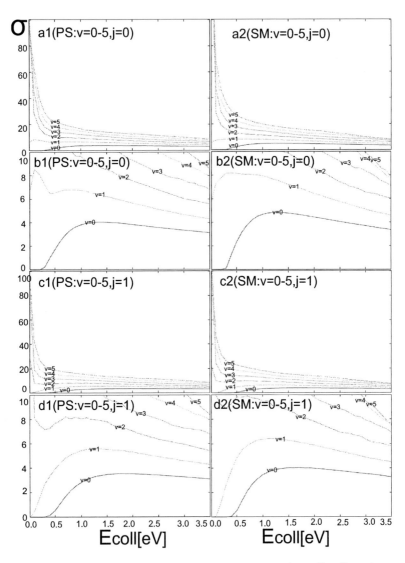

Figure 35: $D^- + H_2(v, j)$ (CT): The total reaction cross sections σ [in 10^{-16} cm^2] as a function of collision energy E_{coll} using PS-PES and SM-PES. Plot $b1(b2, d1, d2)$ is an enlargement of plot $a1(a2, c1, c2)$.

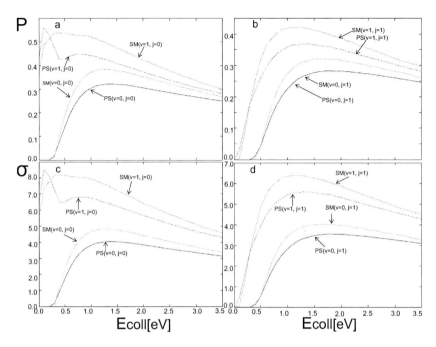

Figure 36: $D^- + H_2(v{=}0{-}1, j{=}0{-}1)$ (CT): Comparison of reaction cross sections σ [in 10^{-16} cm^2] and reaction probabilities P as a function of collision energy E_{coll} for different potential energy surfaces (SM, PS).

for $D^- + H_2(v{=}0, j{=}1)$ (see panels $a1$ and $a2$ in Fig. 35), i.e., σ increases with an increase in E_{coll}. For $E_{coll} = 1.30$ eV and PS-PES σ reaches its maximum ($\sigma_{max} = 5.597{*}10^{-16}$ cm^2). The corresponding maximum is $6.407{*}10^{-16}$ cm^2 at $E_{coll} = 1.10$ eV for SM-PES. These maxima are lower than those for $D^- + H_2(v{=}1, j{=}0)$.

For $D^- + H_2(v{=}3{-}5, j{=}0{-}1)$, the characteristics of the reaction cross sections are similar. The reaction cross sections are quite high ($\sigma > 100{*}10^{-16}$ cm^2) at $E_{coll} = 0.01$ eV; with an increase in the collision energy from 0.01 to 0.5 eV, σ decreases rapidly. When the collision energies are in the range of 0.5–3.5 eV, σ decreases slowly with an increase in E_{coll}. It should be pointed out, however, that the reaction cross sections for $D^- + H_2(v{=}2, j{=}1)$ are much lower than those for $D^- + H_2(v{=}2, j{=}0)$ if the collision energies are in the range of 0.01–0.5 eV.

In order to understand the differences in the reaction cross sections for PS-PES and SM-PES, reaction cross sections and reaction probabilities for $D^- + H_2(v{=}0{-}1, j{=}0{-}1)$ are shown in the same figure (see Fig. 36). One can see that the reaction barriers are the same using PS-PES and SM-PES, but for the same initial rovibrational state, the reaction cross sections using

SM-PES are higher than those using PS-PES. In the case of the H_2 reactant using the same initial vibrational state, the reaction cross sections for $D^- + H_2(v, j=0)$ are higher than those ones for $D^- + H_2(v, j=1)$.

5.4.2 Wave packet calculations for $D^- + H_2(v, j) \rightarrow HD(v', j') + H^-$

Total and state-to-state reaction probabilities (P_{tot}, $P_{v'}$) for the ion-neutral molecule collisions $D^- + H_2(v=0\text{--}1, j=0) \rightarrow HD(v', j') + H^-$ with total angular momentum $J = 0$ were investigated by wave packets (WPs) using the real wave packet code of S. Gray [149]. For the total angular momenta $J \neq 0$, the reaction probabilities, including all total angular momenta up to $J_{max} = 50$, were calculated in the total energy range $E_{tot} = 0.3\text{--}2.0$ eV. In addition, integral cross sections are presented for $E_{tot} = 0.3\text{--}2.0$ eV. The parameters used in the calculations are the same as in the case of $H^- + H_2(v, j) \rightarrow H_2(v', j') + H^-$, as shown in Tab. 16.

Time-dependent quantum dynamics of $D^- + H_2(v, j) \rightarrow HD(v', j') + H^-$ *for angular momentum $J = 0$:*
The ion-neutral molecule collisions $D^- + H_2(v=0\text{--}1, j=0) \rightarrow HD(v', j') + H^-$ were investigated using time-dependent wave packets (WPs) for PS-PES, SM-PES, and AY-PES. In order to obtain accurate results, the total reaction probabilities were calculated with reactant Jacobi coordinates (RC) and product Jacobi coordinates (PC), as shown in the left panels of Fig. 37. The total reaction probabilities are similar for $E_{tot} = 0.3\text{--}2.5$ eV.

Right panels in Fig. 37 show the total and vibrational resolved reaction probabilities (P_{tot}, $P_{v'}$ for $D^- + H_2(v=0\text{--}1, j=0) \rightarrow HD(v', j') + H^-$ and total angular momentum $J = 0$. The total reaction energies (E_{tot}) are chosen in the range of 0.3–3.5 eV. All these calculations are determined using product Jacobi coordinates (PC). It is evident from panels $b1$, $b2$, and $b3$ in Fig. 37 that in the case of the $H_2(v=0, j=0)$ reactant the total reaction probabilities steeply increased beginning from a total energy of 0.48 eV to 1.0 eV followed by a slow increase up to ~1.16 eV with a maximum value of ~86%, ~91%, and ~91% using PS-PES, SM-PES, and AY-PES. For $E_{tot} = 1.2\text{--}2.7$ eV the total reaction probabilities slowly decrease with an increase in E_{tot}. The state-to-state calculations show that at low total energies ($0.3 < E_{tot} < 0.6$ eV) the HD product have the final vibrational state $v'=0$; the other vibrational states of the HD product become populated with an increase in E_{tot}. In the case of $E_{tot} < 1.5$ eV the main product is $HD(v'=0)$, the other vibrational states are minor products.

In the case of the $H_2(v=1, j=0)$ reactant (see panels $c1$, $c2$, $c3$, $d1$, $d2$, and $d3$ in Fig. 37), accurate reaction probabilities near the threshold are difficult to get and lead to a sharp, numerically erroneous peak using product Jacobi coordinates (PC); therefore, there are some differences between the results using reactant Jacobi coordinates (RC) and those using product Jacobi coordinates (PC) at low total energies ($0.7 < E_{tot} < 1.5$ eV). The vibrational distributions of the products are shown in panels $d1$, $d2$, and $d3$ in Fig. 37. It is evident that at low energies the main product is $HD(v'=1)$. The $HD(v'=0)$ product is the minor one at low total energies. For $E_{tot} > 1.1$ eV, products with final higher vibrational states ($v' \geq 2$) can be found.

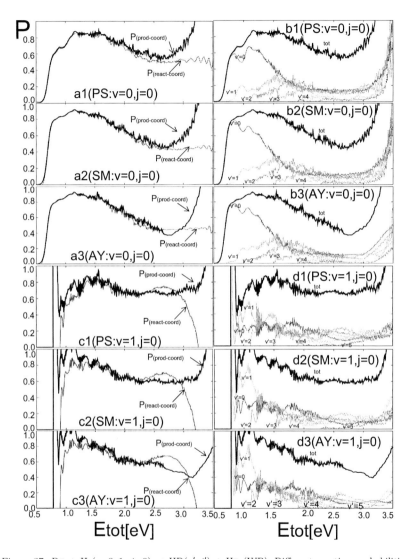

Figure 37: $D^- + H_2(v{=}0{-}1,\ j{=}0) \to HD(v',j') + H^-$ (WP): Different reaction probabilities P ($P_{tot}, P_{v'}$: see abbreviation tot or $v' = \cdots$) as a function of total energy E_{tot} for various initial vibrational states v and two different Jacobi coordinate systems (RC and PC) using the potential energy surfaces SM, PS, and AY ($P_{v'} = \sum_{j'} P(v',j')$, $P_{tot} = \sum_{v'} P_{v'}$).

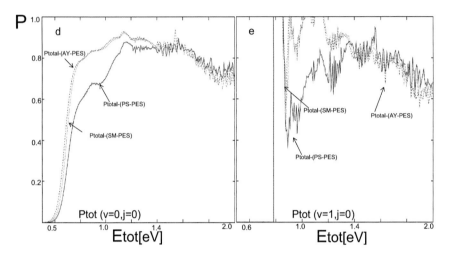

Figure 38: D⁻ + H₂(v=0–1, j=0) (WP): Comparison of total reaction probabilities P as a function of total energy E_{tot} for the different potential energy surfaces (PS-PES, SM-PES, and AY-PES) using product coordinates (PC).

Fig. 38 shows the comparison of the total reaction probability results using PS-PES, SM-PES, and AY-PES for D⁻ + H₂(v=0–1, j=0). For $E_{tot} = 0.3$–2.0 eV the reaction probabilities using SM-PES are similar to those using AY-PES. For $E_{tot} < 1.4$ eV the total reaction probabilities using PS-PES are much lower than those using SM-PES and AY-PES.

Time-dependent quantum dynamics of D⁻ + H₂(v, j) → HD(v', j') + H⁻ *for angular momenta* $J \neq 0$:
Fig. 39 shows the total reaction probabilities using wave packets (WPs) for D⁻ + H₂(v=0, j=0) and the total angular momenta $J = 0, 5, 10, 15, 20, 25, 30, 40$ for SM-PES and PS-PES. The total energies were chosen in the range of 0.3–2.0 eV. It is evident that the trend of the total reaction probabilities and the maxima are quite different between SM-PES and PS-PES. For low total angular momenta J, such as $J = 0$–5, the trend of the total reaction probabilities is similar. The maximum of the reaction probability using SM-PES is larger than that using PS-PES. For the total angular momenta $J = 20$ and 25 and at high total energy, the deviations of the reaction probabilities are large.

In Fig. 40 the reaction cross sections σ [10⁻¹⁶ cm²] are presented for PS-PES and SM-PES for collisions of D⁻ + H₂(v=0, j=0) with WPs. σ includes the summation of reaction probabilities for all J values up to 50 in the total energy range of 0.3–2.0 eV. These results are compared with the results of the quasi-classical trajectories (QCTs), experimental results, and other theoretical results. For the case of using the same potential energy surface, σ calculated with

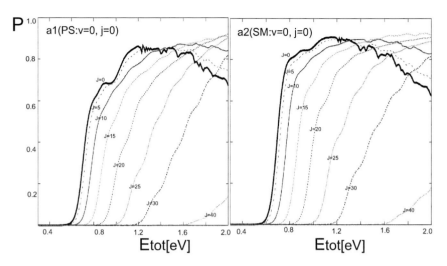

Figure 39: $D^- + H_2(v{=}0,\, j{=}0)$ (WP): Total reaction probabilities P (using reactant coordinates (RC)) as a function of total energy E_{tot} for different total angular momenta $J = 0,\, 5,$ $10,\, 15,\, 20,\, 25,\, 30,\, 40.$

QCTs is larger compared to WPs, and in the case of using the same method, σ for SM-PES is slightly higher than for PS-PES. In the case of WPs, the deviations between SM-PES and PS-PES are not greater than $0.5{*}10^{-16}$ cm^2. It shows that at lower total energies ($0.5 < E_{tot}$ < 0.7 eV), σ of QCTs using PS-PES and SM-PES (see lines b and d in Fig. 40) is in good agreement with the experimental results by Haufler et al. [67] (see points g in Fig. 40). For E_{tot} in the range of $0.8 < E_{tot} < 1.2$ eV, the present results of using WPs (see line a and c in Fig. 40) are closer to the experimental results of Haufler et al. [67] than using QCTs results. For $E_{tot} > 1.2$ eV, the present results are much higher than the experimental results. There are two reasons for these differences. First, higher rotational excited states might be included in the experiment, while in our calculations, we only considered the pure initial rovibrational state $v = 0,\, j = 0$. If the other excited rotational states are taken into account, the computed cross section should be closer to the experimental results. Secondly, for total energies greater than 1.2 eV, the coupling to higher electronically excited states (i.e., ionization to DH$_2$ plus one free electron) has to be included in the dynamics, which would explain why, experimentally, a decrease in the cross section occurs before 2.0 eV.

For E_{tot} in the range of 0.5–1.3 eV, the present WP results of σ using PS-PES are similar to σ of Yao et al. [153] using PS-PES (see line f in Fig. 40). For $E_{tot} > 1.4$ eV, the present results of σ are slightly higher than those of Yao et al. [153]. This may be because the integral cross sections are also calculated up to $\omega = 9$ (J=0-65) in the same way as in case of H$^-$ + D$_2$ in Ref. [153]. Panda et al. [150] investigated this reaction using the centrifugal sudden

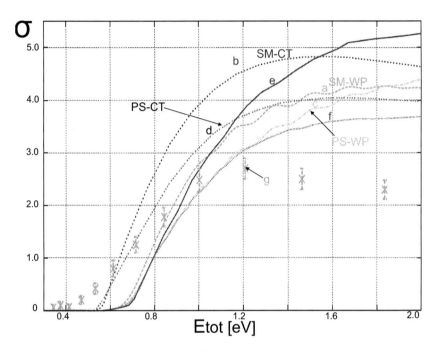

Figure 40: Reaction cross sections σ [10^{-16} cm^2] for $D^- + H_2(v=0, j=0)$: Comparison of theoretical (quantum and classical) and experimental results. a: σ based on wave packets (WP) using SM-PES($J = 0$–50). b: QCT calculations using SM-PES. c: σ based on wave packets (WP) using PS-PES($J = 0$–50). d: QCT calculations using PS-PES. e: σ of Panda et al. (2005) [150] using PS-PES. f: σ of Yao et al. (2006) [153] using PS-PES. g: Haufler et al. (1997) [67].

approximation for PS-PES (see line e in Fig. 40). From Fig. 40, one can see that in the low energy range ($0.5 < E_{tot} < 0.8$ eV), the present results of σ are similar to those of Panda et al. [150]. For $E_{tot} > 1.6$ eV, the results of Panda et al. [150] are significantly higher than the present results. This might be related to the neglect of Coriolis coupling.

5.5 $\mathbf{D^-} + \mathbf{D_2}(v,j)$

The $D^- + D_2(v,j) \rightarrow D_2(v',j') + D^-$ reaction was theoretically investigated by time dependent wave packets (WPs) and quasi-classical trajectories (QCT). The energy dependence of total, state-to-state reaction probabilities and reaction cross sections for the title reactions were calculated.

5.5.1   Quasi-classical trajectory investigations

$D^- + D_2(v{=}0{-}5,\ j{=}0{-}1) \rightarrow D_2(v',j') + D^-$ was studied using the CTAMYM program for the potential energy surfaces of Stärck and Meyer (SM-PES) [50] and Panda and Sathyamurthy (PS-PES) [53].

Initial parameter determinations:
The maximum impact parameter b_{max} was calculated in the same way as in the case of $H^- + H_2(v,j)$. The results are listed in Tab. 24.

Table 24: The maximum impact parameters b_{max} for $D^- + D_2(v{=}0{-}5,\ j{=}0{-}1)$.

v	$v=0$	$v=1$	$v=2$	$v=3$	$v=4$	$v=5$
b_{max} [Å]	1.6	1.9	6.7	10.2	10.3	10.4

Reaction probabilities:
The total and vibrationally resolved reaction probabilities (P_{tot}, $P_{v'}$) were calculated for the $D_2(v{=}0{-}5,\ j{=}0{-}1)$ reactants using SM-PES and PS-PES. The results are shown in Figs. 41 and 42. All the left panels in these figures show the reaction probabilities using PS-PES, while the right panels show the reaction probabilities using SM-PES. Important parts are enlarged in the same panel. The maximum impact parameter b_{max} [Å] used in the calculations are shown. As shown in panels $a1$, $a2$, $b1$, $b2$, $c1$, $c2$, $d1$, and $d2$ in Fig. 41 the trend of the total reaction probability is similar for the $D_2(v{=}0{-}1,\ j{=}0{-}1)$ reactants, i.e., for $E_{coll} > E_{bar}$, the total reaction probabilities P_{tot} steeply increase, reaching a maximum with an increase in E_{coll}. However, for different initial states, the maxima are different. The maximum reaction probabilities using SM-PES are higher than those using PS-PES. In case of using the same potential energy surface, the maximum reaction probabilities for the $D_2(j{=}1)$ reactant are lower than those for the $D_2(j{=}0)$ reactant.
The state-to-state reaction probabilities $P_{v'}$ for $D^- + D_2(v{=}0,\ j{=}0{-}1) \rightarrow D_2(v',j') + D^-$ have the same features as those for $H^- + H_2(v{=}0,\ j{=}0{-}1)$, $D^- + H_2(v{=}0,\ j{=}0{-}1)$, and $H^- + D_2(v{=}0,\ j{=}0{-}1)$, as discussed in previous chapters. At low energies, the highest reaction probabilities are found for the $D_2(v'{=}0)$ product. The $D_2(v'{=}1)$ product will be found only at $E_{coll} > 0.5$ eV. If one wants to find the $D_2(v'{=}2{-}4)$ products, the initial collision energy should be greater than 1.0, 1.2, and 1.5 eV.

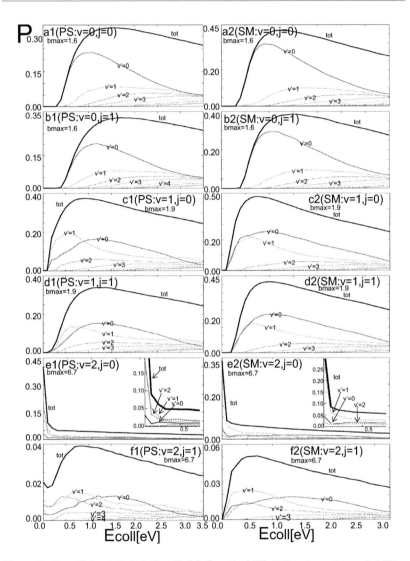

Figure 41: $D^- + D_2(v=0\text{–}2,\ j=0\text{–}1) \to D_2(v',j') + D^-$ (CT): Different reaction probabilities P (P_{tot}, $P_{v'}$: see abbreviation *tot* or $v' = \cdot\cdot$) as a function of collision energy E_{coll} for various rovibrational initial states v,j and different impact parameters b_{max} using the potential energy surfaces SM and PS ($P_{v'} = \sum_{j'} P(v',j')$, $P_{tot} = \sum_{v'} P_{v'}$).

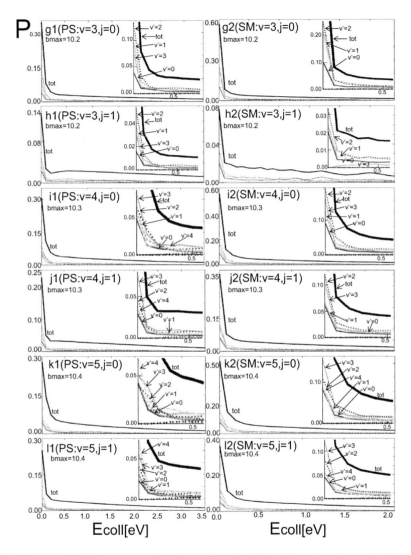

Figure 42: $D^- + D_2(v{=}3{-}5, \, j{=}0{-}1) \rightarrow D_2(v', j') + D^-$ (CT): Different reaction probabilities P (P_{tot}, $P_{v'}$: see abbreviation tot or $v' = \cdot \cdot$) as a function of collision energy E_{coll} for various rovibrational initial states v, j and different impact parameters b_{max} using the potential energy surfaces SM and PS ($P_{v'} = \sum_{j'} P(v', j')$, $P_{tot} = \sum_{v'} P_{v'}$).

For the $D_2(v{=}1, j{=}0)$ reactant (see panels $c1$ and $c2$ in Fig. 41), the state-to-state reaction probabilities using PS-PES show that, at low energies, $(E_{coll}{\leq}0.8$ eV) the main product is $D_2(v'{=}1)$. For $E_{coll} > 0.8$ eV, the $D_2(v'{=}0)$ product is strongly favored. The features of the state-to-state reaction probabilities using SM-PES are similar for PS-PES, i.e., for $E_{coll} < 0.4$ eV the main product is $D_2(v'{=}1)$. For $E_{coll} > 0.4$ eV $D_2(v'{=}0)$ is the most important product. In the case of the $D_2(v{=}1, j{=}1)$ reactant (see panels $d1$ and $d2$ in Fig 41), the features of the vibrational state distributions for the products are different between SM-PES and PS-PES. For PS-PES and $E_{coll} = 0.2–0.8$ eV, the main product is $D_2(v'{=}1)$; for $E_{coll} = 0.8–3.0$ eV, the $D_2(v'{=}0)$ product is strongly favored; for even higher energies ($E_{coll} > 3.0$ eV), again the $D_2(v'{=}1)$ product is the most important one. For SM-PES and $E_{coll} = 0.2–0.6$ eV the reaction probabilities of the $D_2(v'{=}0)$ and $D_2(v'{=}1)$ products are nearly the same. For E_{coll} in the range of $0.2–0.4$ eV, the reaction probabilities for the $D_2(v'{=}0)$ product are slightly higher than those for $D_2(v'{=}1)$; for $E_{coll} = 0.4–0.6$ eV, the $D_2(v'{=}1)$ product is slightly dominating; for E_{coll} in the range of $0.6–3.5$ eV, the $D_2(v'{=}0)$ product is strongly favored.

It should be pointed out that in the case of the $D_2(v{=}2)$ reactant, the total reaction probabilities, using PS-PES, are higher than those using SM-PES, and the total reaction probabilities for the $D_2(v{=}2, j{=}1)$ reactant are much lower than those for the $D_2(v{=}2, j{=}0)$ reactant. This is totally different from $H^- + H_2(v{=}2, j{=}0–1)$, $H^- + D_2(v{=}2, j{=}0–1)$, and $D^- + H_2(v{=}2, j{=}0–1)$. The state-to-state reaction probabilities show that using PS-PES and low collision energies ($E_{coll} < 0.3$ eV), the main products are $D_2(v'{=}1)$ and $D_2(v'{=}2)$. $D_2(v'{=}0)$ is a minor product. In the case of using SM-PES and low collision energies ($E_{coll} < 0.5$ eV), the most important product is $D_2(v'{=}1)$.

Fig. 42 shows the total and vibrationally resolved reaction probabilities for the $D_2(v{=}3–5, j{=}0–1)$ reactants using PS-PES and SM-PES. In these reactions, the total reaction probabilities are high at $E_{coll} = 0.01$ eV; then, the reaction probabilities steeply decrease with an increase in E_{coll}. For $E_{coll} > 0.3$ eV, the total reaction probabilities decrease slowly with an increase in E_{coll}. For PS-PES, the main product is $D_2(v'{=}v{-}1)$. For SM-PES with the $D_2(v{=}3, j{=}0–1)$ reactant, the main product is $D_2(v'{=}2)$. At low energies with the $D_2(v{>}3)$ reactants, the product D_2 is mainly produced with the final vibrational state $v'{=}v{-}2$.

Here, we should point out that the total reaction probabilities for $D^- + D_2(v{=}3, j{=}1)$ using SM-PES are rather low. At higher energies, the corresponding reaction cross sections are even lower than those for the $D_2(v{=}0–2)$ reactants (we will discuss this in the following section).

Reaction cross sections:

The reaction cross sections σ $[10^{-16}$ cm$^2]$ for the $D^- + D_2(v{=}0–5, j{=}0–1)$ reaction were investigated with quasi-classical trajectories (QCT) using Eq. (58) as shown in Fig. 43. Collision energies were chosen in the range of $0.01–3.5$ eV. σ for different initial vibrational states using PS-PES is shown in the left panels of Fig. 43. In these panels, $b1$ and $d1$ are the enlargements of panels $a1$ and $c1$. σ using SM-PES is shown in the right panels of Fig. 43. In these panels, $b2$ and $d2$ are the enlargements of panels $a2$ and $c2$.

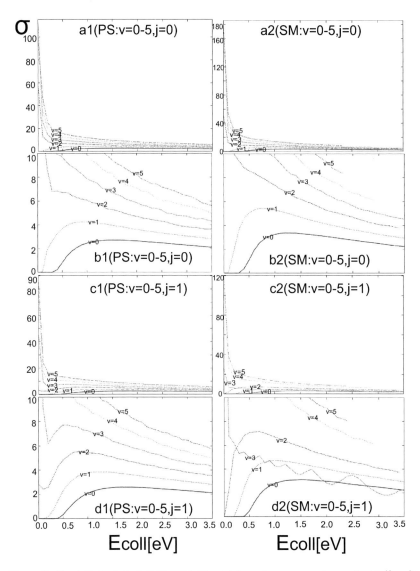

Figure 43: $D^- + D_2(v{=}0{-}5,\ j{=}0{-}1)$ (CT): The total reaction cross sections σ [in 10^{-16} cm^2] as a function of collision energy E_{coll} using PS-PES and SM-PES. Plot $b1(b2,d1,d2)$ is an enlargement of plot $a1(a2,c1,c2)$.

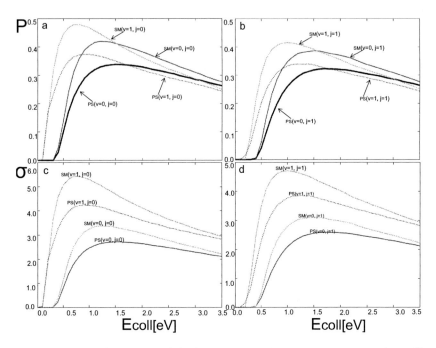

Figure 44: $D^- + D_2(v=0–1, j=0–1)$ (CT): Comparison of reaction cross sections σ [in 10^{-16} cm^2] and reaction probabilities P as a function of collision energy E_{coll} for different potential energy surfaces (SM, PS).

For the $D_2(v=0, j=0)$ reactant, σ increases dramatically in the case of $E_{coll} > E_{bar}$. For $E_{coll} = 1.50/1.20$ eV σ reaches its maximum at $2.72/3.38*10^{-16}$ cm^2 using PS-PES/SM-PES, and then σ slowly decreases with an increase in E_{coll}. A similar result is obtained with $H^- + H_2(v=0, j=0)$, $D^- + H_2(v=0, j=0)$, and $H^- + D_2(v=0, j=0)$ at low collision energies. σ for SM-PES is much higher than for PS-PES. The trend of the reaction cross section for the $D_2(v=0, j=1)$ reactant is similar to those one for the $D_2(v=0, j=0)$ reactant. The corresponding maxima are 2.59 and $3.11*10^{-16}$ cm^2, which are much lower than those for $D^- + D_2(v=0, j=0)$.

For the $D_2(v=1)$ reactant, the features of total reaction cross sections are similar to those for the $D_2(v=0)$ reactant. The maximum reaction cross sections are 4.24 and $3.85*10^{-16}$ cm^2 for the $D_2(v=1,i \; j=0)$ and $D_2(v=1, j=1)$ reactants using PS-PES. Using SM-PES, the corresponding σ_{max} are 5.43 and $4.72*10^{-16}$ cm^2 for the $D_2(v=1, j=0)$ and $D_2(v=1, j=1)$ reactants.

The trend for σ is very different between $D^- + D_2(v=2, j=0)$ and $D^- + D_2(v=2, j=1)$. In the case of the $D_2(v=2, j=0)$ reactant, the reaction cross sections are high ($56.575*10^{-16}$ cm^2 using PS-PES, $50.477*10^{-16}$ cm^2 using SM-PES) at $E_{coll} = 0.01$ eV; beyond these maxima, the reaction cross sections decrease with an increase in E_{coll}. For the $D_2(v=2, j=1)$ reactant, however, the reaction cross sections are low ($2.797*10^{-16}$ cm^2 using PS-PES, $0.004*10^{-16}$ cm^2 using SM-PES) at $E_{coll} = 0.01$ eV; then the cross sections increase with an increase in E_{coll}, and the reaction cross sections reach its maxima 5.51 and $7.16*10^{-16}$ cm^2 at $E_{coll} = 0.8/0.7$ eV using PS-PES/SM-PES.

In the case of the $D_2(v=3-5, j=0-1)$ reactant, the characteristics of the reaction cross sections are similar. σ is high at $E_{coll} = 0.01$ eV. For $E_{coll} = 0.01-0.5$ eV σ decreases rapidly with an increase in E_{coll}. For $E_{coll} > 0.5$ eV the reaction cross sections decrease slowly with an increase in collision energies.

Here, we should point out that, as shown in panel $d2$ in Fig. 43, σ is quite low for $E_{coll} > 0.2$ eV for the $D_2(v=3, j=1)$ reactant using SM-PES. At higher collision energies σ is even lower than for the $D_2(v=0, j=1)$ reactant. This is quite an abnormal feature in all our calculations.

5.5.2 Wave packet calculations for $D^- + D_2(v,j) \rightarrow D_2(v',j') + D^-$

The ion-molecule collisions $D^- + D_2(v,j) \rightarrow D_2(v',j') + D^-$ for total angular momenta $J = 0$ and $J \neq 0$ were investigated using wave packet (WP) program [149]. The parameters used in the calculations are the same as in the case of $H^- + H_2(v,j) \rightarrow H_2(v',j') + H^-$, as shown in Tab. 16. The total integral reaction cross sections σ were calculated.

Time-dependent quantum dynamics of $D^- + D_2(v,j) \rightarrow D_2(v',j') + D^-$ *for total angular momentum* $J = 0$:
The total and vibrationally resolved reaction probabilities for $D^- + D_2(v=0-1, j=0) \rightarrow D_2(v',j') + D^-$ for total angular momentum $J = 0$ were calculated using the three potential energy surfaces PS, SM and AY. The total energies are in the range of 0.3–3.5 eV. Details are shown in Fig. 45. In order to get accurate results, one calculates the total reaction probabilities using reactant Jacobi coordinates (RC) and product Jacobi coordinates (PC). As shown in the left panels of Fig. 45 for the total energies $E_{tot} < 2.5$ eV, the total reaction probabilities using RC-coordinates are similar to those using PC-coordinates.

The right panels in Fig. 45 show the energy dependent total and state-to-state reaction probabilities (P_{tot}, $P_{v'}$) using product Jacobi coordinates (PC) for PS-PES, SM-PES, and AY-PES. For the $D_2(v=0, j=0)$ reactant (as shown at panels $b1$, $b2$, and $b3$ in Fig. 45), the total reaction probabilities increase immediately at $E_{tot} = 0.5-0.7$ eV with maxima of ~50%, ~61%, and ~61% for PS-PES, SM-PES, and AY-PES; for PS-PES, there is a slow increase up to 1.6 eV with a magnitude of ~80%, but for SM-PES and AY-PES and $E_{tot} = 1.3$ eV, the maximum is ~83%. In the case of the $D_2(v=0, j=0)$ reactant, the trend features for the vibrationally resolved reaction probabilities are similar for PS-PES, SM-PES, and AY-PES. At low energies ($E_{tot} < 1.5$ eV), the final vibrational state $v' = 0$ is strongly favored in the D_2 product. $D_2(v'=1)$, as one of minor products, can be found at $E_{tot} > 0.8$ eV.

In the case of the $D_2(v=1)$ reactant accurate total reaction probabilities, which were calculated for product Jacobi coordinates (PC) near the threshold are difficult to obtain and lead to a sharp, numerically erroneous peak. Panels $c1$, $c2$, and $c3$ in Fig. 45 for $E_{tot} = 0.75-2.5$ eV

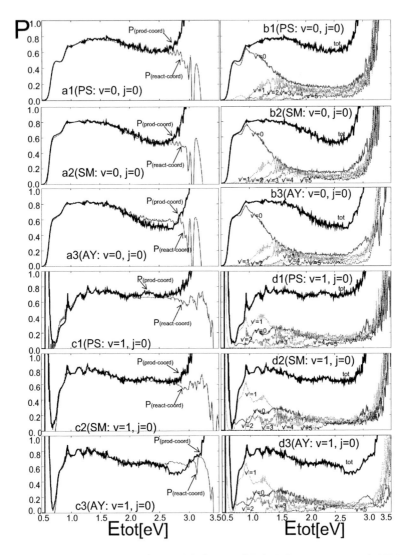

Figure 45: $D^- + D_2(v{=}0{-}1,\ j{=}0) \rightarrow D_2(v', j') + D^-$ (WP): Different reaction probabilities P (P_{tot}, $P_{v'}$: see abbreviation tot or $v' = \cdots$) as a function of total energy E_{tot} for various vibrational initial states v and two different Jacobi coordinate systems (RC and PC) using the potential energy surfaces SM, PS, and AY ($P_{v'} = \sum_{j'} P(v', j')$, $P_{tot} = \sum_{v'} P_{v'}$).

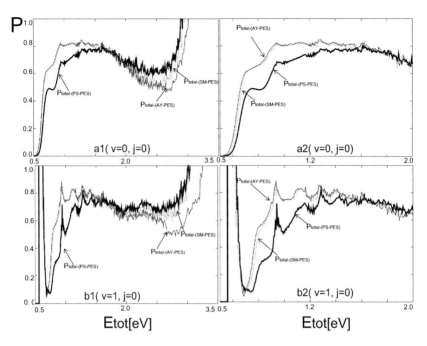

Figure 46: $D^- + D_2(v=0–1, j=0)$ (WP): Comparison of total reaction probabilities P as a function of total energy E_{tot} for the different potential energy surfaces (PS, SM, and AY) using product coordinates (PC).

show accurate total reaction probabilities using product Jacobi coordinates (PC). The total reaction probabilities steeply increase at $E_{tot} \approx$ 0.7–0.93 eV with a sharp maximum of ~70%, ~85%, and ~90% for PS-PES, SM-PES, and AY-PES, followed by a slow decrease in total reaction probabilities with an increase in E_{tot}. For E_{tot} in the range of 0.75–1.5 eV, the D_2 product mainly has the final vibrational state $v' = 1$. For $1.5 < E_{tot} < 2.0$ eV, the $D_2(v'=0)$ product is slightly dominating.

For $D^- + D_2(v=0–1, j=0)$ and $E_{tot} > 2.5$ eV, the reaction probabilities increase with an increase in E_{tot}, and for even higher energies ($E_{tot} > 3.0$ eV), the total reaction probabilities are even higher than 100%.

Fig. 46 shows a comparison of the total reaction probabilities using PS-PES, SM-PES, and AY-PES. The right two panels are enlargements of the left two panels. From these panels, it is evident that for $E_{tot} < 1.6$ eV the total reaction probabilities for $D^- + D_2(v=0–1, j=0) \rightarrow D_2(v', j') + D^-$ using SM-PES and AY-PES are similar, but in case of PS-PES these are much lower.

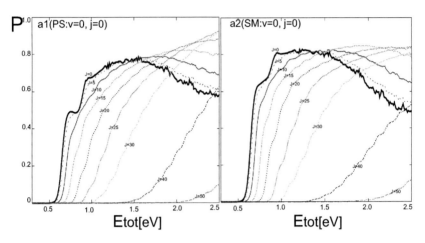

Figure 47: $D^- + D_2(v=0, j=0)$ (WP): Total reaction probabilities P (using reactant coordinates (RC)) as a function of total energy E_{tot} for different total angular momenta $J = 0, 5, 10, 15, 20, 25, 30, 40, 50$.

Time-dependent quantum dynamics of $D^- + D_2(v=0, j=0) \rightarrow D_2(v', j') + D^-$ *for total angular momenta $J \neq 0$:*

Fig. 47 presents total reaction probabilities for $D^- + D_2(v=0, j=0)$ for the total angular momenta $J = 0, 5, 10, 15, 20, 25, 30, 40, 50$ using wave packets (WPs) program for PS-PES (left panel $a1$) and SM-PES (right panel $a2$). The total energies were chosen in the range of 0.3-2.5 eV. Fig. 47 shows the trend of the total reaction probabilities; the maxima for SM-PES are different from those for PS-PES. In the case of total angular momenta $J = 0, 5, 10, 15$, the maxima of reaction probabilities using SM-PES are higher than those using PS-PES, but the trend in the reaction probabilities is similar. In the case of total angular momenta $J > 15$ and $E_{tot} > 1.5$ eV, the total reaction probabilities using PS-PES increase faster with an increase in E_{tot} compared to SM-PES. For the total angular momenta $J = 25$ and 30 and $E_{tot} = 2.5$ eV, the total reaction probabilities using PS-PES are much higher than those using SM-PES. In Fig. 48, the total cross section σ, which was calculated with wave packets (WPs) (J up to 50) and quasi-classical trajectories (QCT) (using the CTAMYM program), is presented using PS-PES and SM-PES for $D^- + D_2(v=0, j=0)$. Fig. 48 shows that the reaction barriers using WPs and QCT are similar for PS-PES and SM-PES. Using SM-PES the curve for σ using QCTs is similar to the one using WPs, i.e., for $E_{tot} > E_{bar}$ σ increases with an increase in E_{tot}. The reaction cross sections reach their maxima $3.4/3.2*10^{-16}$ cm^2 at $E_{tot} = 1.4/1.6$ eV for QCTs and WPs. Beyond the maximum σ decreases slowly with an increase in total energy. In the entire energy range, σ using QCTs is higher than in the case of using WPs. For PS-PES

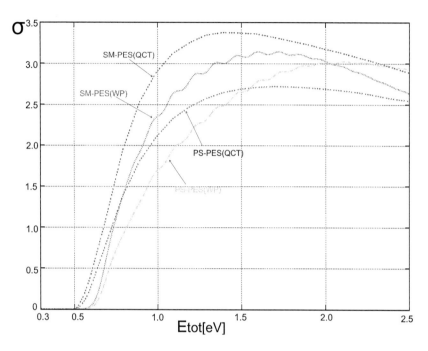

Figure 48: Reaction cross sections σ [in 10^{-16} cm^2] for D^- + $D_2(v=0,\ j=0)$: Comparison of quantum and classical calculations using the potential surfaces SM and PS.(WP: quantum wave packets, QCT: quasi-classical trajectories.)

and $E_{tot} < 1.6$ eV, σ using QCTs is larger than in the case of using WPs, but at higher total energies ($E_{tot} > 1.6$ eV), σ using WPs is much higher than in case of using QCTs. If this reaction is investigated classically for $E_{tot} = 0.3$–2.0 eV, using PS-PES σ is lower than in case of using SM-PES. The reason has been discussed in H^- + H_2 chapter. Similar results are are seen for quantum investigations. In case of WPs and high total energies ($E_{tot} > 2.1$ eV), σ calculated for PS-PES is higher than σ calculated for SM-PES.

5.6   H⁻ + HD(v, j) and D⁻ + HD(v, j)

The collisions of H⁻ and D⁻ ions with HD(v=0–1, j=0) were theoretically investigated by quasi-classical trajectories (QCT) (CTAMYM program). The energy dependence of reaction probabilities and reaction cross sections were calculated in this work.

The maximum impact parameters b_{max} were calculated in the same way as in the case of H⁻ + H$_2$(v, j). The results are listed in Tab. 25. The initial internal energies, turning points, and

Table 25: The maximum impact parameters b_{max} for H⁻ + HD(v=0–1, j=0) and D⁻ + HD(v=0–1, j=0).

v	H⁻ + HD(v=0)	H⁻ + HD(v=1)	D⁻ + HD(v=0)	D⁻ + HD(v=1)
b_{max} [Å]	1.6	2.2	1.3	2.2

vibrational half-periods for HD using PS-PES and SM-PES were investigated. The results are shown in Tabs. 26 and 27.

Table 26: Internal energies, turning points, and vibrational half-periods of HD for different vibrational (v) and rotational (j) states (using SM-PES).

v	j	$E_{v,j}$[cm⁻¹]	r_- [Å]	r_+ [Å]	τ[fs]
0	0	1881.954	0.640233	0.87299	0.4497670
0	1	1971.038	0.641036	0.87391	0.4501622
1	0	5505.399	0.580393	0.99077	0.4711845
1	1	5590.735	0.581163	0.99175	0.4715536
2	0	8960.776	0.545307	1.08572	0.4946460
2	1	9042.390	0.546058	1.08676	0.4952437
3	0	12246.302	0.520085	1.17313	0.5210709
3	1	12324.196	0.520823	1.17424	0.5216948
4	0	15360.070	0.500482	1.25770	0.5509608
4	1	15434.228	0.501212	1.25888	0.5515796
5	0	18300.107	0.484596	1.34185	0.5846259
5	1	18370.498	0.485320	1.34311	0.5852997

5.6.1   Reaction probabilities for the collision of D⁻ and H⁻ with HD(v=0–1, j=0)

The total reaction probabilities and different product reaction probabilities for D⁻ + HD(v=0–1, j=0) were calculated using PS-PES and SM-PES. The results are shown in Fig. 49. Panel A in Fig. 49 shows the reaction probabilities P_A for D⁻ + HD(v=0, j=0) → D$_2$(v', j') + H⁻ and P_B for D⁻ + HD(v=0, j=0) → DH(v', j') + D⁻. P_{tot} ($P_{tot} = P_A + P_B$) is the total reaction probability. From panel A in Fig. 49, one can see that the total reaction probability starts to increase for total energy E_{tot} ~0.5 eV; beyond this energy, the total reaction probability steeply increases up to a maximum at E_{tot} ~1.5 eV. Beyond the maximum, the total reaction

Table 27: Internal energies, turning points, and vibrational half-periods of HD for different vibrational (v) and rotational (j) states (using PS-PES).

v	j	$E_{v,j}[\text{cm}^{-1}]$	r_- [Å]	r_+ [Å]	$\tau[fs]$
0	0	1887.800	0.641389	0.87361	0.4485961
0	1	1976.608	0.642182	0.87453	0.4489100
1	0	5514.388	0.581841	0.99214	0.4715738
1	1	5599.287	0.582604	0.99313	0.4720051
2	0	8962.651	0.546893	1.08779	0.4962249
2	1	9043.729	0.547641	1.08884	0.4967419
3	0	12237.645	0.521673	1.17577	0.5232049
3	1	12314.920	0.522412	1.17687	0.5238182
4	0	15342.305	0.501950	1.26069	0.5531688
4	1	15415.733	0.502686	1.26186	0.5537849
5	0	18278.913	0.485836	1.34492	0.5865315
5	1	18348.399	0.486572	1.34615	0.5872678

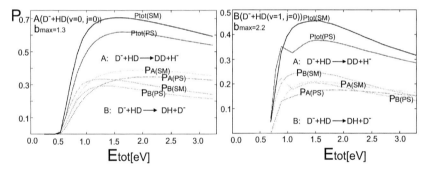

Figure 49: A: D^- + HD(v=0–1, j=0) $\rightarrow$ DH(v', j') + D^- and B: D^- + HD(v=0–1, j=0) $\rightarrow$ $D_2(v', j')$ + H^- (CT): Different reaction probabilities P (P_{tot}, P_A, and P_B) as a function of total energy E_{tot} for various vibrational initial states v and different impact parameters b_{max} using the potential energy surfaces SM and PS (P_{tot}(SM) = P_A(SM) + P_B(SM), P_{tot}(PS) = P_A(PS) + P_B(PS)).

probability decreases slowly with an increase in E_{tot}. The results for P_{tot} using SM-PES are higher than those using PS-PES in the entire energy range. For E_{tot} in the range of 0.5–1.1 eV, the reaction probabilities P_B(SM, PS) are higher than P_A(SM, PS). This means that, in the given energy range, the main product is HD(v', j') + D^-. With an increase in E_{tot} the $D_2(v', j')$ + H^- product is dominant and more important for D^- + HD(v=0, j=0). For E_{tot} > 3.0 eV, the reaction probabilities P_A(SM, PS) are nearly double the reaction probabilities P_B(SM, PS). Panel B in Fig. 49 shows the reaction probabilities P_A for D^- + HD(v=1, j=0)

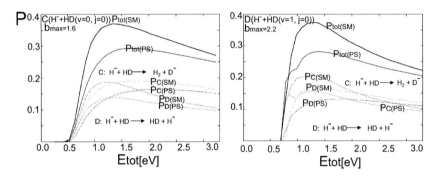

Figure 50: C: H^- + $HD(v=0\text{--}1, j=0) \rightarrow HD(v',j') + H^-$ and D: H^- + $HD(v=0\text{--}1, j=0) \rightarrow$ $H_2(v',j') + D^-$ (CT): Different reaction probabilities P (P_{tot}, P_C, and P_D) as a function of total energy E_{tot} for various vibrational initial states v and different impact parameters b_{max} using the potential energy surfaces SM and PS (P_{tot}(SM) = P_C(SM) + P_D(SM), P_{tot}(PS) = P_C(PS) + P_D(PS)).

$\rightarrow D_2(v',j') + H^-$ and P_B for D^- + $HD(v=1, j=0) \rightarrow DH(v',j') + D^-$. In the case of using SM-PES and $E_{tot} \approx 0.7$ eV, P_{tot}(SM) increases immediately up to its maximum of $\approx 45\%$. In the case that E_{tot} are in the range of 1.5–3.5 eV, P_{tot} decreases with increasing E_{tot}. In the case of using PS-PES, P_{tot}(PS) has a sharp maximum at $E_{tot} = 0.9$ eV. For $E_{tot} = 1.0$ eV, P_{tot} reaches a minimum. The reaction probability increases again with increasing E_{tot}; after the second maximum, P_{tot} decreases slowly with an increase in E_{tot}. For $E_{tot} < 2.5$ eV, the reaction probability for the $DH(v',j') + D^-$ product is much higher than that for the $D_2(v',j')$ + H^- product. For $E_{tot} > 2.5$ eV, the reaction probabilities are nearly the same for the two products.

Fig. 50 shows the reaction probabilities of the total and different products for H^- + $HD(v=0$–1, $j=0$) using PS-PES and SM-PES. Panel C in Fig. 50 shows the total and different product reaction probabilities for the $HD(v=0, j=0)$ reactant. In panel C, one can see that the total reaction probabilities P_{tot}(SM) and P_{tot}(PS) start to increase for $E_{tot} > 0.5$ eV. At $E_{tot} \approx 1.5$ eV, the total reaction probability reaches its maximum P_{tot}(SM) $\approx 37\%$ and P_{tot}(PS) $\approx 29\%$ using SM-PES and PS-PES. For $E_{tot} > 1.5$ eV, the total reaction probabilities decrease with an increase in E_{tot}. In the low energy range $0.5 < E_{tot} < 1.2$ eV, the reaction probabilities of channel $H_2(v',j') + D^-$ are slightly higher than those of the channel $HD(v',j') + H^-$. For $E_{tot} > 1.5$ eV, the reaction probabilities of the channel $H_2(v',j') + D^-$ are much lower than those of the channel $HD(v',j') + H^-$. The total and different channel reaction probabilities for H^- + $HD(v=1, j=0)$ are shown in panel D of Fig. 50. In the case of $E_{tot} > E_{bar}$, the total reaction probability steeply increases with an increase in E_{tot}. At $E_{tot} = 1.3$ eV, the total reaction probabilities reach their maxima P_{tot}(SM) $\approx 37\%$ and P_{tot}(PS) $\approx 28\%$ using SM-PES and PS-PES. For E_{tot} in the range of 0.6–2.0 eV, the main products are $HD(v',j') + H^-$. For

$E_{tot} > 2.0$ eV, the reaction probabilities of the channel HD(v', j') + H⁻ and H₂(v', j') + D⁻ are similar.

5.6.2 Reaction cross sections for the collisions of D⁻ and H⁻ ions with HD(v=0–1, j=0)

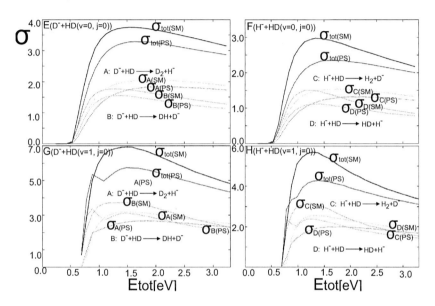

Figure 51: D⁻ + HD(v=0–1, j=0) and H⁻ + HD(v=0–1, j=0) (CT): Reaction cross sections σ [in 10^{-16} cm²] as a function of total energy E_{tot} calculated with quasi-classical trajectories for the potential energy surfaces SM and PS.

The reaction cross sections σ for D⁻ + HD(v=0–1, j=0) and H⁻ + HD(v=0–1, j=0) were investigated with quasi-classical trajectories (QCT) using SM-PES and PS-PES. The results are shown in Fig. 51. The product branching ratios were calculated using the same method. The results are shown in Fig. 52. Panels C and D in Fig. 52 are enlargements of panels A and B. Total energies were chosen in the range of 0.0–3.5 eV. Panels E and G in Fig. 51 show σ for D⁻ + HD(v=0–1, j=0) using PS-PES and SM-PES. In the case of the HD(v=0, j=0) reactant and $E_{tot} > E_{bar}$ σ steeply increases up to its maximum, then σ decreases slowly with an increase in E_{tot}. σ for SM-PES is higher than for PS-PES in the whole energy range. For $E_{tot} < 1.2$ eV, σ for the DH(v', j') + D⁻ product is larger than for the D₂(v', j') + H⁻ product. From panel A in Fig. 52, one can see that when case E_{tot} are in the range of 0.53–1.2 eV, the ratio σ_{HD}/σ_{D_2} is greater than 1.0. It should be pointed out that at the fixed total energy of $E_{tot} = 0.53$ eV, the value of the product ratio σ_{HD}/σ_{D_2} is 90.9 (using PS-PES) and 39.5 (using

107

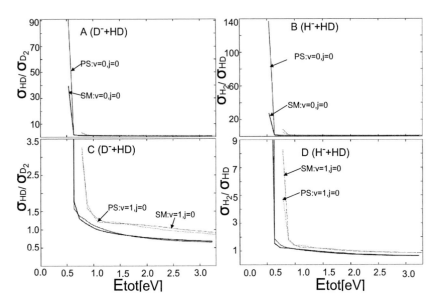

Figure 52: $D^- + HD(v{=}0{-}1, j{=}0)$ and $H^- + HD(v{=}0{-}1, j{=}0)$ (CT): Product branching ratios calculated with quasi-classical trajectories for the potential energy surfaces SM and PS.

SM-PES). This means that, in the given total energy range, the HD product is remarkably preferred over the D_2 product. The product ratio decreases sharply with an increase in E_{tot}. At higher energies, the product ratio σ_{HD}/σ_{D_2} is nearly equal to 0.6. In the case of the HD($v{=}1$, $j{=}0$) reactant (as shown at panel G in Fig. 51 and panel C in Fig. 52) and $E_{tot} < 2.5$ eV, the $HD(v', j') + D^-$ product is favored. For $E_{tot} > 2.5$ eV, the product ratio σ_{HD}/σ_{D_2} is nearly equal to one. These two products are of equal importance in this given energy range.

Panels F and H in Fig. 51 show σ for $H^- + HD(v{=}0{-}1, j{=}0)$ using PS-PES and SM-PES. The characteristics of σ are similar to those of $D^- + HD(v{=}0{-}1, j{=}0)$. In the case of the HD($v{=}0$, $j{=}0$) reactant (see panel F in Fig. 51 and right panels in Fig. 52) and $E_{tot} < 1.3$ eV, the main product is $H_2(v', j') + D^-$. In this energy range, the product ratio σ_{H_2}/σ_{HD} is greater than 1.0. For $E_{tot} = 0.53$ eV, the product ratio is 137.5 and 27.9 using PS-PES and SM-PES. With an increase in E_{tot}, the product ratio decreases sharply. For $E_{tot} > 1.4$ eV, the product ratio is lower than 1.0. At even higher total energies of $E_{tot} > 3.0$ eV, the product ratio nearly equals 0.6, meaning that in the higher total energy range, the main product is $HD(v', j') + H^-$. In the case of the HD($v{=}1$, $j{=}0$) reactant and at a fixed total energy of 0.78 eV, the product ratio σ_{H_2}/σ_{HD} is 4.8 and 8.3 using PS-PES and SM-PES (see the panel D in Fig. 52). The product ratio decreases sharply with an increase in E_{tot}. For $E_{tot} > 2.1$ eV, the product ratio σ_{H_2}/σ_{HD} is less than 1.0. In this total energy range, the $HD(v', j') + H^-$

product is slightly dominates.

6 Appendix

6.1 Associated Legendre functions

The first associated Legendre functions are listed below:

$$P_1^1(\cos(\theta)) = (1 - \cos^2(\theta))^{\frac{1}{2}} = \sin(\theta) \tag{142}$$

$$P_2^1(\cos(\theta)) = 3\cos(1 - \cos^2(\theta))^{\frac{1}{2}} = 3\cos(\theta)\sin(\theta) \tag{143}$$

$$P_2^2(\cos(\theta)) = 3(1 - \cos^2(\theta)) = 3\sin^2(\theta) \tag{144}$$

$$P_3^1(\cos(\theta)) = \frac{3}{2}(5\cos^2(\theta) - 1)(1 - \cos^2\theta)^{\frac{1}{2}} = \frac{3}{2}(5\cos^2(\theta) - 1)\sin(\theta) \tag{145}$$

$$P_3^2(\cos(\theta)) = 15\cos(\theta)(1 - \cos^2(\theta)) = 15\cos(\theta)\sin^2(\theta) \tag{146}$$

$$P_3^3(\cos(\theta)) = 15\cos(\theta)(1 - \cos^2(\theta))^{\frac{3}{2}} = 15\sin^3(\theta) \tag{147}$$

6.2 Legendre polynomial

Legendre polynomials are defined by

$$P_n(x) = \frac{d^n}{2^n n! dx^n}[(x^2 - 1)^n] \ . \tag{148}$$

The first five Legendre polynomials are

$$P_0(x) = 1,$$
$$P_1(x) = x,$$
$$P_2(x) = \frac{1}{2}(3x^2 - 1),$$
$$P_3(x) = \frac{1}{2}(5x^3 - 3x),$$
$$P_4(x) = \frac{1}{8}(35x^4 - 30x^2 + 3). \tag{149}$$

6.3 Fortran code for calculation of turning point and vibrational period of a diatomic molecule

```
      PROGRAM TURNING
C
C      PROGRAM TO CALCULATE TURNING POINTS OF A MOLECULE
C      PROGRAM TO CALCULATE THE HALF–PERIOD  OF A MOLECULE
C
      IMPLICIT DOUBLE PRECISION(A–H,O–Z)
```

```
        COMMON /EVIB/ EVIBROT,IROT
        WRITE(6,*) 'V    J    EV,J(cm-1)    TURNIN(AU)  TURNOU(AU)   T (AU)
     1    T(FS)'
        NUE=0.0
        READ(5,*) NMAX,NROTMAX
        WRITE(6,*)  'NMAX,NROTMAX'
        WRITE(6,*)  NMAX,NROTMAX
        READ(5,*) XMIN,XMAX
        WRITE(6,*)  'XMIN,XMAX'
        WRITE(6,*)  XMIN,XMAX
        WRITE(16,*) 'IVIB,   IROT,   XMAXIN  ,YMINN'
        DO 100 N=1,NMAX                                    !SM-H3-MINUS HAVE 14 STATES.
        READ(5,*) IVIB
 101    DO 10 M=1,NROTMAX              !J MAX DELETE TO 32
        READ(5,*,ERR=20,END=30) IROT,EVIBROT
        IF(IROT.GT.0) GO TO 102
 C      FIND MINIMUM AND CALCULATE XMAXIN
        CALL ZERO(XMIN,XMAXIN,YMINN,IROT)
        WRITE(16,*) IVIB,IROT,XMAXIN,YMINN
 102    CONTINUE
        CALL   BISECTION(XMIN,XMAXIN,TURNIN)
        CALL   BISECTION(XMAXIN,XMAX,TURNOU)
        XMAXIN=TURNOU
        CALL VIBPER(TURNIN,TURNOU,HALFPE)
        WRITE(6,'(2I6,F12.3,4F13.8)')
       *IVIB,IROT,EVIBROT,TURNIN,TURNOU,HALFPE,HALFPE*2.41888436505D-2
        CALL FLUSH(6)
   10   CONTINUE
   20   CONTINUE
        WRITE(6,*)
        IVIB=IROT
        GOTO 101
 100    CONTINUE
  30    CONTINUE
        END
 C
 C      THE FOLLOWING PROGRAM CALCULATES THE TURNING POINT
 C      TURNIN IS THE INNER TURNING POINT
 C      TURNOU IS THE OUTER TURNING POINT
 C
```

```
       SUBROUTINE BISECTION(XMINO,XMAXO,XMID)
       INPLICIT DOUBLE PRECISION(A-H,O-Z)
C      COMMON FWERTE,TURN,I,J
       EXTERNAL FWERT
       XMIN=XMINO
       XMAX=XMAXO
       ITMAX=1000
       ACCU=1.d-7
C      MIDDLE POINT
       XMID=(XMIN+XMAX)*0.5D0
       FMIN=FWERT(XMIN)
       FMAX=FWERT(XMAX)
       FMID=FWERT(XMID)
       IF(FMIN*FMAX.FT.0.0d0) THEN
       WRITE(6,*) 'XMIN,XMAX,XMID'
       WRITE(6,*) XMIN,XMAX,XMID
       WRITE(6,*) 'FMIN,FMAX,FMID'
       WRITE(6,*) FMIN,FMAX,FMID
       STOP ' MIST'
       ENDIF
       ITER=0
100    PMIN=FMIN*FMID
C      PMAX=FMAX*FMID
       ITER=ITER+1
       IF(PMIN.LT.0.0d0) THEN
       XMAX=XMID
       FMAX=FMID
       ELSE
       XMIN=XMID
       FMIN=FMID
       ENDIF
       XMID=(XMIN+XMAX)*0.5D0
       FMID=FWERT(XMID)
       XX=ABS((XMAX-XMID)/XMID)
       IF(XX.LE.1.D-7.AND.FMID.LE.0.d0) RETURN
       IF(ITER.GT.ITMAX) WRITE(6,55) ITMAX
       GOTO 100
55     FORMAT(//' SORRY, FAILED TO CONVERGE IN', I6, ' TRIES.'/)
       RETURN
       END
```

```
C
C THE FOLLOWING SUBROUTINE IS A FUNCTION GET THE POTENTIAL ENERGY
C
      FUNCTION FWERT (X)
      IMPLICIT DOUBLE PRECISION A-H,O-Z)
      COMMON /EVIB/ EVIBROT,JROT
      DIMENSION RR(3),GRAD(3)
      AMA=1.00727646688D0          !ATOM H
      AMB=1.00727646688D0          !ATOM H
      AMA=AMA*1822.888427181D0
      AMB=AMB*1822.888427181D0
C 1AMU=1.6605402D-27KG
C      AMA=AMA*1.6605402D-27
C      AMB=AMB*1.6605402D-27                       !CHANGE AMU TO KG
      UAB=(AMA*AMB)/(AMA+AMB)
C      ENERGIES IN HARTREE
      EFACT=219474.6313705D0                 ! TRANSFORM CM-1 TO AU(HF)
      RR(1)=X
      RR(2)=3000.D0
      RR(3)=3000.D0
      KEY=1
      KION=0
      CALL POTH3M(RR,E,GRAD,KEY,KION,VH3)
      POT=E*1.D-3
C NOW POTENTIAL INCLUDING ROTATIONAL PART
      POTROT=(POT + JROT*(JROT+1)/(2.D0*UAB*X*X))*EFACT
      FWERT=POTROT-EVIBROT
      RETURN
      END
      SUBROUTINE VIBPER(TURNIN,TURNOU,HALFPE)
      IMPLICIT DOUBLE PRECISION(A-H,O-Z)
      COMMON /EVIB/ EVIBROT,JROT
      DIMENSION RR(3),GRAD(3)
      EFACT=219474.6313705D0
C INPUT DATA UNITS IS AU(LENGTH),AMU(MASS),AU(ENERGY)
C  I SHOULD CHANGE THE LENGTH TO METER THE ENERGY TO JOULE,THE MASS TO KG.
      J=JROT
      ENJ=EVIBROT/EFACT
      KEY=1
      KION=0
```

```
      RR(2)=3000.d0
      RR(3)=3000.d0
      EPS1=1.D-7
      EPS2=1.D-6
      AMA=1.00727646688D0
      AMB=1.00727646688d0
      AMA=AMA*1822.888427181D0
      AMB=AMB*1822.888427181D0
      UAB=(AMA*AMB)/(AMA+AMB)
      DD0=0.D0
      DD1=0.D0
      DD2=0.D0
      DD3=0.D0
      N1=1.D-2/EPS1
      R=TURNIN-EPS1
      DO 1=1,N1                      !THIS IS THE FIRST PART FROM RIN TO 1.D-2
      R=R+EPS1
      RR(1)=R
      CALL POTH3M(RR,E,GRAD,KEY,KION,VH3)
C     !H. STCK W. MEYER, CHEM. PHYS. 176, 83 (1993)
      VAB=E*1.D-3
      EROT=J*(J+1)/(2.D0*UAB*R**2)
      DD=EPS1*(1.D0/DSQRT(ENJ-VAB-EROT))
      DD1=DD1+DD
      END DO
      R=TURNOU-1.D-2
      DO 1=1,N1-1      !THIS IS THE SECOND PART FROM ROUTER TO ROUTER-1.D-2
      R=R+EPS1
      RR(1)=R
      CALL POTH3M(RR,E,GRAD,KEY,KION,VH3)
      VAB=E*1.D-3
      EROT=J*(J+1)/(2.D0*UAB*R**2)
      DD=EPS1*(1.D0/DSQRT(ENJ-VAB-EROT))
      DD2=DD2+DD
      END DO
      R=TURNIN+1.D-2-EPS2
      N2=(TURNOU-TURNIN-2.D-2)/EPS2
      DO L=1,N2 !THIS IS THE THIRD PART FROM RIN+1.D-2 TO ROUTER-1.D-2
      R=R+EPS2
      RR(1)=R
```

```
      CALL POTH3M(RR,E,GRAD,KEY,KION,VH3)
      VAB=E*1.D-3
      EROT=J*(J+1)/(2.D0*UAB*R**2)
      DD=EPS2*(1.D0/DSQRT(ENJ-VAB-EROT))
      DD3=DD3+DD
      END DO
      DD0=DD1+DD2+DD3
      HALFPE=DSQRT(UAB/2.D0)*DD0
      RETURN
      END
C
C CALCULATE THE LOWEST ENERGY POINT.
C
      SUBROUTINE ZERO(XMINA,XMAXIN,YMINN,JROT)
      IMPLICIT DOUBLE PRECISION(A-H,O-Z)
      DIMENSION RR(3),GRAD(3)
      XMIN=XMINA
      EPS=1.D-5
      DEREPS=1.D-5
      DEREPS2=1.D-1
      PASOEPS=1.D-4
      AMA=1.00727646688D0
      AMB=1.00727646688D0
      AMA=AMA*1822.888427181D0
      AMB=AMB*1822.888427181D0
C 1AMU=1.6605402D-27KG
C      AMA=AMA*1.6605402D-27
C      AMB=AMB*1.6605402D-27                          !CHANGE AMU TO KG
      UAB=(AMA*AMB)/(AMA+AMB)
      KEY=1
      KION=0
      RR(2)=300.D0
      RR(3)=300.D0
      YMIN=1.D11
      RR(1)=XMIN
      CALL POTH3M(RR,E,GRAD,KEY,KION,VH3)
      Y=E*1.D-3
      DO 20 J=1,999999
      RR(1)=RR(1)+EPS
      CALL POTH3M(RR,E,GRAD,KEY,KION,VH3)
```

```
Y2=E*1.D-3+JROT*(JROT+1)/(2.D0*UAB*RR(1)*RR(1))
IF(Y2.LT.YMIN) THEN
YMIN=Y2
XMIN=RR(1)
ENDIF
RR(1)=RR(1)-2.D0*EPS
CALL POTH3M(RR,E,GRAD,KEY,KION,VH3)
Y3=E*1.D-3+JROT*(JROT+1)/(2.D0*UAB*RR(1)*RR(1))
IF(Y3.LT.YMIN) THEN
YMIN=Y3
XMIN=RR(1)
ENDIF
RR(1)=RR(1)+EPS
DER1=(Y2-Y3)/(2.D0*EPS)
DERDER=(Y2-Y)/EPS
DERIZQ=(Y-Y3)/EPS
DER2=(Y2+Y3-2.D0*Y)/(EPS*EPS)
PASO0=-Y/DER1
PASO1=-DER1/DER2
IF(DABS(DER1).GT.DEREPS2) THEN
PASO=PASO0
ELSE
PASO=PASO1
ENDIF
RR(1)=RR(1)+PASO
CALL POTH3M(RR,E,GRAD,KEY,KION,VH3)
Y=E*1.D-3+JROT*(JROT+1)/(2.D0*UAB*RR(1)*RR(1))
IF(Y.LT.YMIN) THEN
YMIN=Y
XMIN=RR(1)
ELSEIF(DABS(PASO).GT.PASOEPS) THEN
DO 100 K=1,2
RR(1)=RR(1)-PASO
PASO=PASO*0.1D0
RR(1)=RR(1)+PASO
CALL POTH3M(RR,E,GRAD,KEY,KION,VH3)
Y=E*1.D-3+JROT*(JROT+1)/(2.D0*UAB*RR(1)*RR(1))
IF(Y.LT.YMIN) THEN
YMIN=Y
XMIN=RR(1)
```

```
      ENDIF
100   CONTINUE
      ENDIF
      RR(1)=XMIN
      Y=YMIN
      DIF=0.D0
      DERI=0.D0
      DIF=DIF+DABS(PP0-RR(1))
      PP0=RR(1)
      DER1=DABS(DER1)
CWANG      IF(DIF.LT.PASOEPS.AND.DER1.LT.DEREPS) THEN
      IF(DIF.LT.PASOEPS) THEN
      YMINN=YMIN
      XMAXIN=XMIN
      RETURN
      ENDIF
 20   CONTINUE
      PRINT*,' WARNING: NUMBER OF ITERATIONS EXCEDED'
      RETURN
      END
C
C INPUT FOR H2
C
C    999,32            : NMAX, NROTMAX
C    0.1d0,100.d0,1.4d0          : XMAXIN,XMIN,XMAX
C    0
C    0     2170.466
C
C OUTPUT FOR H2 USING SM-PES
C
C V J  EV,J(CM-1) TURNIN(AU) TURNOU(AU)  T (AU)     T(FS)
C 0 0  2170.466   1.19727041 1.67073370  161.5542  3.9078
C
```

References

[1] A. Rau, J. Astophys. Astr. 17, 113 (1996).

[2] J. Hirschfelder, H. Eyring, and N. Rosen, J. Chem. Phys. 4, 121 (1936)

[3] J. Hirschfelder, H. Eyring, and N. Rosen, J. Chem. Phys. 4, 130 (1936).

[4] J. Hirschfelder, H. Diamond, and H. Eyring, J. Chem. Phys. 5, 695 (1937).

[5] A. Farkas, L. Farkas, Proc. Roy. Soc A152, 124 (1935).

[6] J. J. Thomson, Phil. Mag. 24, 225 (1911).

[7] A. J. Dempster, Phil. Mag. 31, 438 (1916).

[8] H. D. Smyth, Rew. Mod. Phys. 3, 347 (1931).

[9] C. A. Coulson, Proc. Camb. Phil. Soc. 31(2), 244 (1935).

[10] I. R. McNab, Adv. Chem. Phys. 89, 1 (1995).

[11] T. Oka, Proc. Natl. Acad. Sci. U.S.A. 103, 12235 (2006).

[12] T. Oka, Philos. Trans. R. Soc. London, Ser. A 364, 2847 (2006).

[13] T. Oka, Phys. Rew. Lett. 45, 531 (1980).

[14] C. M. Lindsay and B. J. McCall, J. Mol. Spectrosc. 210, 60 (2001).

[15] J. L. Gottfried, B. J. McCall, T. Oka, J. Chem. Phys. 118, 10890 (2003).

[16] J. L. Gottfried, Philos, Trans. R. Soc. London, Ser. A 364, 2917 (2006).

[17] A. Carrington, J. Buttenshaw, R. A. Kennedy, Mol. Phys. 45, 753 (1982).

[18] A. Carrington, R. A. Kennedy, J. Chem. Phys. 81, 91 (1984).

[19] A. Carrington, I. R. McNab, Y. D. West, J. Chem. Phys. 98, 1073 (1993).

[20] P.Drossart, J. P. Maillard, J. Caldwell, S. J. Kim, J. K. G. Watson, W. A. Majewski, J. Tennyson, S. Miller, S. K. Atreya, J. T. Clarke, J. H. Waite, Jr., R. Wagener, Nature (London) 340, 539 (1989).

[21] T. R. Geballe, T. Oka, Nature (London) 384, 334 (1996).

[22] B. J. McCall, T. R. Geballe, K. H. Hinkle, T. Oka, Science 279, 1910 (1998).

[23] W. Meyer, P. Botschwina, P. Burton, J. Chem. Phys. 84, 891 (1986).

[24] R. Röhse, W. Kutzelnigg, R. Jaquet, W. Klopper, J. Chem. Phys. 101, 2231 (1994).

[25] W. Cencek, J. Rychlewski, R. Jaquet, W. Kutzelnigg, J. Chem. Phys. 108, 2831 (1998).

[26] R. Jaquet, W. Cencek, W. Kutzelnigg, J. Rychlewski, J. Chem. Phys. 108, 2831 (1998).

[27] R. Jaquet, Spectrochim. Acta, Part A 58, 691 (2002).

[28] B. Dinelli, S. Miller, J. Tennyson, J. Mol. Spectrosc. 163, 71 (1994).

[29] R. Prosmiti, O. L. Polyansky, J. Tennyson, Chem. Phys. Lett. 273, 107 (1997).

[30] O. L. Polyansky, R. Prosmiti, W. Klopper, J. Tennyson, Mol. Phys 98, 261 (2000).

[31] A. Aguado, O. Rocero, C. Tablero, C. Sany, M. Paniagua, J. Chem. Phys. 112, 1240 (2000).

[32] C. Sanz, O. Roncero, C. Tablero, A. Aguado, M. Paniagua, J. Chem. Phys. 114, 2182 (2001).

[33] L. Velilla, B. Lepetit, A. Aguado, J. A. Beswick, M. Paniagua, J. Chem. Phys. 129, 084307 (2008).

[34] R. K. Preston, J. C. Tully, J. Chem. Phys. 54, 4297 (1971).

[35] H. Kamisaka, W. Bian, K. Nobusada, and H. Nakamura, J. Chem. Phys. Vol. 116, 654 (2002)

[36] W. Kutzelnigg, R. Jaquet, Phil. Trans. R. Soc. A. 364, 2855 (2006).

[37] K. N. Crabtree, C. A. Kauffman, B. A. Tom, E. Becka, B. A. McGuire, B. J.McCall, J. Chem. Phys. 134, 194311 (2011).

[38] S. Gómez-Carrasco, L. González-Sánchez, A. Aguado, C. Sanz-Sanz, A. Zanchet, O. Roncero, J. Chem. Phys. 137, 094303 (2012).

[39] L. P. Viegas, A. Alijah, A. J. C. Varandas, J. Chem. Phys. 126, 074309 (2007).

[40] A. J. C. Varandas, in Conical Intersections: Electronic Structure. Dynamics and Spectoscopy, edited by D. Yarkony, H. Köppel, W. Domcke (World Scientific, Singapore, 2004), Chap. 5.

[41] D. Stevenson and J. Hirschfelder, J. Chem. Phys. 5, 933 (1937).

[42] R. E. Hurley, Nucl. Instrum. Methods 118, 307 (1974).

[43] W. Aberth, R. Schnitzer, M. Anbar, Phys. Rev. Lett. 34, 1600 (1975).

[44] R. Schnitzer, R. W. Odom, M. Anbar, J. Chem. Phys. 68, 1489 (1978).

[45] Y. K. Bae, M. J. Coggiola, J. R. Peterson, Phys. Rev. A 29, 2888 (1984).

[46] W. Wang, A. K. Belyaev, Y. Xu, A. Zhu, C. Xiao, X.-F. Yang, Cpem. Phys. Lett. 377, 512 (2003).

[47] R. Golser, H. Gnaser, W. Kutschera, A. Priller, P. Steier, A. Wallner, M. Čížek, J. Horáček, W. Domcke, Phys. Rev. Lett. 94, 223003 (2005).

[48] H. Gnaser, R. Golser, Phys. Rev. A 73, 021202(R) (2006).

[49] W. John, The Science of JET. (2000).

[50] J. Stärck and W. Meyer, Chem. Phys. 176, 83 (1993).

[51] F. Robicheaux, Phys. Rev. A 60, 1706 (1999).

[52] P. R. Bunker and P. Jensen, Molecular Symmetry and Spectroscopy (NRC Research Press, Ottawa, Canada, 1998).

[53] A. N. Panda, N. Sathyamurthy, J. Chem. Phys. 121, 9343 (2004).

[54] M. Ayouz, O. Dulieu, R. Guérout, J. Robert, V. Kokoouline, J. Chem. Phys. 132, 194309 (2010).

[55] A. K. Belyaev, A. S. Tiukanov, Chem. Phys. 220, 43 (1997).

[56] O. K. Kabbaj, F. Volatron, and J.-P. Malrieu, Chem. Phys. Lett. 147, 353 (1988).

[57] A. K. Belyaev, D. T. Colbert, G. C. Groenenboom, and W. H. Miller, Chem. Phys. Lett. 209, 309 (1993).

[58] A. K. Belyaev and A. S. Tiukanov, Chem. Phys. Lett. 302, 65 (1999).

[59] A. K. Belyaev, A. S. Tiukanov, and W. Domcke, Phys. Rev. A 65, 012508 (2001).

[60] A. K. Belyaev, A. S. Tiukanov, and W. Domcke, Phys. Scrip. 80, 048124 (2009).

[61] J. E. E. Muschlitz, T. L. Bailey, and J. H. Simons, J. Chem. Phys. 24, 1202 (1956).

[62] J. E. E. Muschlitz, T. L. Bailey, and J. H. Simons, J. Chem. Phys. 26, 711 (1957).

[63] M. S. Huq, L. D. Doverspike, and R. L. Champion, Phys. Rev. A 27, 2831 (1983).

[64] M. Zimmer and F. Linder, Chem. Phys. Lett. 195, 153 (1992).

[65] M. Zimmer and F. Linder, J. Phys. B: At. Mol. Opt. Phys. 28, 2671 (1995).

[66] H. Müller, Z. Zimmer, and F. Linder, J. Phys. B: At. Mol. Opt. Phys. 29, 4165 (1996).

[67] E. Haufler, S. Schlemmer, and D. Gerlich, J. Phys. Chem. 101, 6441 (1997).

[68] S. Mahapatra, N. Sathyamurthy, S. Kumar, F. A. Gianturco, Chem. Phys. Lett. 241, 223 (1995).

[69] W. H. Ansari, N. Sathyamurthy, Chem. Phys. Lett. 289, 487 (1998).

[70] F. Aguillon, A. K. Belyaev, V. Sidis, M. Sizum, Phys. Chem. Chem. Phys., 2, 3577 (2000).

[71] R, Jaquet, M. Heinen, J. Phys. Chem. A 105, 2738 (2001).

[72] C. Morari, R. Jaquet, J. Phys. Chem. A 109, 3396 (2005).

[73] K. Giri, N. Sathyamurthy, J. Phys. B: At. Mol. Opt. Phys. 39, 4123 (2006).

[74] L. Yao, L. Ju, T. Chu, K. L. Han, Phys. Rev. A 74, 062715 (2006).

[75] K. Giri, N. Sathyamurthy, Chem. Phys. Lett. 444, 23 (2007).

[76] W. Li, M. S. Wang, C. Yang, X. Ma, D. Wang, W. Liu, Chem. Phys. Lett. 445, 125 (2007).

[77] M. Ayouz, R. Lopes, M. Raoult, O. Dulieu, V. Kokoouline, Phys. Rev. A 83, 052712 (2011).

[78] J. Mazur, R. Rubin, J. Chem. Phys. 31, 1395 (1959).

[79] D. Kosloff, R. Kosloff, J. Comput. Phys. 52, 35 (1985).

[80] J. V. Lill, G. A. Parker, J. C. Light, Chem. Phys. Lett. 89, 483 (1982)

[81] J. C. Light, I. P. Hamilton, J. V. Lill, J. Chem. Phys. 82, 1400 (1985)

[82] H. Tal-Ezer, R. Kosloff, J. Chem. Phys. 81, 3967 (1984).

[83] V. Mandelshtam, H. S. Taylor, J. Chem. Phys. 102, 7390 (1995).

[84] V. Mandelshtam, H. S. Taylor, J. Chem. Phys. 103, 2903 (1995).

[85] S. K. Gray, G. G. Balint-Kurti, J. Chem. Phys. 108, 950 (1998).

[86] M. Karplus, R. N. Porter, R. D. Sharma, J. Chem. Phys. 43, 3259 (1965).

[87] R. N. Porter, L. M. Raff, in Dynamics of Molecular Collisions, ed. W. H. Miller, Plenum Press, New York, 1976, part B, p. 1.

[88] R. N. Porter, Annu. Rev. Phys. Chem., 1974, 25, 317.

[89] W. L. Hase, in Dynamics of Molecular Collisions, ed. W. H. Miller, Plenum Press, New York, 1976, part B, p. 121.

[90] D.G. Truhlar and J. T. Muckerman, Reactive Scattering Cross Sections III: Quasiclassical and Semiclassical Methods. Atom-Molecule Collision Theory. p. 505 (1979) Ed: R. B. Bernstein.

[91] M. D. Pattengill and R. B. Bernstein, Reactive Scattering Cross Sections III: Quasiclassical and Semiclassical Methods. Atom-Molecule Collision Theory. p. 359 (1979) Ed: R. B. Bernstein.

[92] L. M. Raff and D. L. Thompson, in The Theory of Chemical Reaction Dynamics, ed. M. Baer, CRC Press, Boca Raton, FL, 1985, vol. III, p. 1.

[93] H. R. Mayne, in Dynamics of Molecules and Chemical Reactions, ed. R. E. Wyatt and J. Z. H. Zhang, Marcel Dekker, New York, 1996, p.589.

[94] T. D. Sewell and D. L. Thompson, Int. Mod. Phys. B 11, 1067 (1997).

[95] T. Yonehara, S. Kato J. Chem. Phys. 117, 11131 (2002).

[96] D. C. Jacobs, R. N. Zare, J. Chem. Phys. 91, 3196 (1989).

[97] J. C. Tully, Nonadiabatic Dynamics. Modern methods for multidimensional dynamics computations in chemistry, ed. D. L. Thompson. (p. 34-72) 1998.

[98] J. C. Tully, Nonadiabatic Processes in Molecualr Collisions, ed. W. H. Miller. (p. 217-267) 1976.

[99] T. S. Chu, A. J. C. Varandas, K. L. Han, Chem. Phys. Lett. 417, 22 (2009).

[100] J. C. Tully, R. K. Preston, J. Chem. Phys. 55, 562 (1971).

[101] M. G. Holliday, J. T. Muckerman, L. Friedman, J. Chem. Phys. 54, 1058 (1971).

[102] J. Krenos, R. Wolfgang, J. Chem. Phys. 52, 5961 (1970).

[103] A. Ichihara, T. Shirai, K. Yokoyama, J. Chem. Phys. 105, 1857 (1996).

[104] G.D. Billing, N. Markovic, Chem. Phys. 209, 377 (1996).

[105] N. Markovic, G.D. Billing, Chem. Phys. Lett. 248, 420 (1996).

[106] N. Markovic, G.D. Billing, Chem. Phys. 191, 247 (1995).

[107] I. Last, M. Gilibert, M. Baer, J. Chem. Phys. 107, 1451 (1997).

[108] V.G. Ushakov, K. Nobusada, V.I. Osherov, Phys. Chem. Chem. Phys. 3, 63 (2001).

[109] T. Takayanagi, Y. Kurosaki, A. Ichihara, J. Chem. Phys. 112, 2615 (2000).

[110] H. Kamisaka, W. Bian, K. Nobusada, H. Nakamura, J. Chem. Phys. 116, 654 (2002).

[111] A. Ichihara, K. Yokoyama, J. Chem. Phys. 103, 2109 (1995).

[112] L.P. Viegas, A.J.C. Varandas, Phys. Rev. A 77, 032505 (2008).

[113] O. Friedrich, A. Alijah, Z.R. Xu, A.J.C. Varandas, Phys. Rev. Lett. 86,1183 (2001).

[114] R.F. Lu, T.S. Chu, K.L. Han, J. Phys. Chem. A 109, 6683 (2005).

[115] T.S. Chu, K.L. Han, J. Phys. Chem. A 109, 2050 (2005).

[116] R. J. Le Roy, Level 8.0: a computer program for Solving the Radial Schrödinger Equation for Bound and Quasibound Levels, University of Waterloo Chemical Physics Research Report CP-663 (2007).

[117] K. J. Johnson, Numerical Methods in Chemistry, Department of Chemistry, University of Pittsburg (1980).

[118] S. S. M. Wong, Computational Methods in Physics and Engineering. p. 383-406 (1992).

[119] R. D. Levine and R. B. Bernstein Molecular Reaction Dynamics and Chemical Reactivity. p. 53 (1987).

[120] J. C. Tully J. Chem. Phys. 93, 1061 (1990).

[121] P. Atkins, J. d. Paula, ATKIN's Physical Chemistry. p. 317 (2002).

[122] R. Kosloff, J. Phys. Chem. 92, 2087 (1988).

[123] S. K. Gray, J. Chem. Phys. 96, 6543 (1992).

[124] W. H. Miller, Annu. Rev. Phys. Chem. 41, 245 (1990).

[125] R. T. Pack, J. Chem. Phys. 60, 633 (1974).

[126] A. M. Arthurs, A. Dalgarno, Proc. R. Soc. Lond. A. 256, 540 (1960).

[127] G. G. Balint-Kurti, R. N. Dixon, and C. C. Marston, Internat. Rev. Phys. Chem. 11, 317 (1992).

[128] A. S. Dickinson, P. R. Certain, J. Chem. Phys. 49, 4209 (1968).

[129] X. Balakrishnan, C. Kalyanaraman, N. Sathyamurthy, Physics Reports, 280, 79 (1997).

[130] R. G. Newton, "Scattering theory of waves and particles", Springer (1982).

[131] D. J. Kouri, T. G. Heil, Y. Shimoni, J. Chem. Phys. 65, 226 (1976).

[132] R. S. Judson, D. J. Kouri, D. Neuhauser, M. Baer, Phys. Rev. A, 42, 351 (1990).

[133] D. Neuhauser, M. Baer, R. S. Judson, D. J. Kouri, Comp. Phys. Commun., 63, 460 (1991).

[134] G. G. Balint-Kurti, R. N. Dixon, C. C. Marston, J. Chem. Soc. Faraday Trans., 86, 1741 (1990).

[135] A. R. Offer, G. G. Balint-Kurti, J. Chem. Phys. 101, 10416 (1994).

[136] A. J. Meijer, E. M. Goldfield, S. K. Gray, G. G. Balint-Kurti, Chem. Phys. Lett., 293, 270 (1998).

[137] MOLPRO is a package of *ab initio* programs written by H.-J. Werner and P. J. Knowles with contributions from R. D. Amos *et al.*

[138] W. Meyer, J. Chem. Phys. 58, 1017 (1973).

[139] Z. H. Li, A. W. Jasper, D. A. Bonhommeau, and D. G. Truhlar, ANT 07, University of Minnesota, Minneapolis, 2007.

[140] D. G. Truhlar, J. W. Duff, N. C. Blais, J. C. Tully, B. C. Garrent, J. Chem. Phys. 77, 764 (1982).

[141] A. W. Jasper, D. G. Truhlar, J. Chem. Phys. Vol. 127, 194306 (2007).

[142] A. W. Jasper, S. N. Stechmann, D. G. Truhlar, J. Chem. Phys. 116, 5424 (2002).

[143] V. G. Ushakov, K. Nobusada, and V. I. Osherov, Phys. Chem. Chem. Phys., Vol. 3, 63 (2001).

[144] G. C. Schatz, J. Phys. Chem. 100, 12839 (1996).

[145] S. L. Mielke, B. C. Garrett, and K. A. Peterson, J. Chem. Phys. 116, 4142 (2002).

[146] H. Song, D. Dai, G. Wu, C. C. Wang, S. A. Harich, M. Y. Hayes, X. Wang, D. Gerlich, X. Yang, R. T. Skodje, J. Chem. Phys. 123, 074314 (2005).

[147] D. G. Hopper, QCPE 11, 248 (1974).

[148] W. L. Hase, R. J. Duchovic, X. Hu, A. Komornicki, K.F. Lim, D. H. Lu, G. H. Peslherbe, K. N. Swamy, S. R. V. Linde, A. Varandas, H. B. Wang, and R. J. Wolf, June, 1996.

[149] Real wave packet code of S. Gray. See ref [85].

[150] A. N. Panda, K. Giri, N. Sathyamurthy, J. Phys. Chem. A 109,2057 (2005).

[151] W. L. Li, M. S. Wang, Mol. Phys., 105, 2329 (2007).

[152] P. G. Jambrina, F. J. Aoiz, N. Bulut, S. C. Smith, G. G. Balint-Kurti, M. Hankel, Phys. Chem. Chem. Phys., 12, 1102 (2010).

[153] L. Yao, L. P. Ju, T. S. Chu, K. L. Han, Phys. Rev. A. 74, 062715 (2006).